RESTORING OLD BUILDINGS FOR CONTEMPORARY USES

RESTORING OLD BUILDINGS FOR CONTEMPORARY USES

AN AMERICAN SOURCEBOOK FOR ARCHITECTS AND PRESERVATIONISTS

WILLIAM C. SHOPSIN, AIA

WHITNEY LIBRARY OF DESIGN
an imprint of Watson-Guptill Publications/New York

First published 1986 in New York by the Whitney Library of Design
an imprint of Watson-Guptill Publications,
a division of Billboard Publications, Inc.,
1515 Broadway, New York, N.Y. 10036

Library of Congress Cataloging-in-Publication Data

Shopsin, William C.
Restoring old buildings for contemporary uses.

Bibliography: p.
Includes index.
1. Architecture—United States—Conservation and restoration—Handbooks, manuals, etc. I. Title.
NA106.S56 1986 720'.28'8 86-1655
ISBN 0-8230-7425-0

Distributed in the United Kingdom by Phaidon Press Ltd., Littlegate House, St. Ebbe's St., Oxford

Manufactured in U.S.A.

First Printing, 1986

1 2 3 4 5 6 7 8 9 / 91 90 89 88 87 86

Senior Editor: Julia Moore
Associate Editor: Victoria Craven
Designer: Jay Anning
Production Manager: Ellen Greene
Set in 10 point ITC Bookman Light

ILLUSTRATION CREDITS

Douglas Howard Adil: 1-1 (and chapter opening), 7-17, 7-41, 7-42, back cover photograph

Attia & Perkins: 2-9

Patricia Layman Bazelon: Chenango County Courthouse (photographs of exterior and ceiling detail)

Arne Bystrom, FAIA: Seattle Garden Center (before photograph of exterior, interior photographs, plan drawings and elevations)

The Canadian Inventory of Historic Buildings, Parks Canada: 3-1

Zev Daniels: Olivet Building

The Ehrenkrantz Group, P.C.: 7-4 (and chapter opening), 7-11, 7-12, One Church Street

John Glick: The Bank Center

Warren Gran: 3-5

Tod Henkels, Stephen B. Jacobs & Associates: Portico Place

Nicholas Holmes, Jr., FAIA: Henderson House

Lester Katz: 8-7

Mendel-Mesick-Cohen-Waite Architects: 6-1, 6-2, 6-3, 6-7, 6-8, 7-1, 7-3, 7-5, 7-6, 7-7, 7-8, 7-9, 7-13, 7-14, 7-30, 7-31, 7-33, 7-34, 7-39, 7-40, 7-50, 7-52, 7-53, 7-54, 7-55, 8-2, 8-6, 8-11, 8-12, 8-13, 8-14, 9-2, 9-3, 9-4, 9-7, 9-8, Chenango County Courthouse, Canandigua National Bank

Marc Neuhof: 8-15

Steve Rosenthal: St. Charles Meeting House (photographs)

John A. Sharratt Associates: St. Charles Meeting House (plan drawings)

Susannah Falk Shopsin: 1-5, 1-6, 1-7, 1-8, 2-2, 2-3, 2-6, 2-7, 2-10, 2-11, 3-4, 4-1, 4-3, 4-4 (and chapter opening), 4-5, 4-6, 6-6, 6-9, 6-10, 7-2, 7-10, 7-18, 7-19, 7-22, 7-24, 7-25, 7-26, 7-27, 7-28, 7-44, 7-49, 7-51, 9-5 (and chapter opening), 9-6, Seattle Garden Center (after photograph of exterior)

William C. Shopsin: 1-2, 1-3, 1-4, 2-1, 2-4, 2-5 (and chapter opening), 2-12, 3-2, 6-4 (and chapter opening), 6-5, 7-15, 7-16, 7-20, 7-21, 7-32, 7-35, 7-36, 7-37, 7-38, 7-43, 7-45, 7-46, 7-47, 7-48, 8-3, 8-10, 9-1

Simon Thoresen & Associates (now Sculley, Thoresen & Linard): Whitefield and Beechwood (and chapter opening)

Ralph P. Turcotte: Daniel Low & Company (front cover photograph)

Steven Zane: 2-8, 3-3 (and chapter opening), 4-2, 7-23, 7-29, 8-1, 8-5, 8-8 (and chapter opening), 8-9.

To my students, who've been asking for a basic text for a long time. And to my family, who has been patient and supportive while I wrote it.

ACKNOWLEDGMENTS

Only an author can appreciate the number of people who are actively involved in transforming a topic into a text. Organizing mountains of material and minutae into a readable resource is a truly complex enterprise. During the five-year gestation of this book, I have relied heavily on the editorial staff of the Whitney Library of Design.

Stephen Kliment, FAIA, former Executive Editor at Whitney, recognized the need for a basic historic preservation text and asked me to undertake the project. Former Editor Susan Davis patiently organized my material and provided encouragement that kept me going. Julia Moore, Whitney Senior Editor, transformed my manuscript into a book, assisted by Associate Editor Victoria Craven, who supported me through the logistics of production. Barbara Klinger produced the checklists and organized the bibliography and resource sections.

A number of professional colleagues have been most supportive in sharing their experience and in providing plans and illustrations of their projects. These represent the regional diversity of North American needs and styles: an Alabama mansion restored by Nick Holmes, Jr.; two grand country estates in New York State by Simon Thoreson, Sean Sculley, and Dimitri Linard; a small commercial structure in Seattle by Arne Bystrom; recycled churches in New York City by Steve Jacobs, and in Boston by John Sharratt Associates; and a small shopping mall in Pittsburgh by Mike Marcu. Denis Kuhn of the Ehrenkrantz Group contributed the project in Nashville. David Bergholz of the Allegheny Conference on Community Development gave me valuable background information on downtown Pittsburgh.

I want particularly to acknowledge the cooperation and support received from architects Mendel-Mesick-Cohen-Waite of Albany, New York, whose extensive and well-documented historic preservation practice has generated many of the hard-to-find kinds of illustrations of work in process that are so essential to a book like this one. They also supplied information for two projects in Chapter 5, the Canandaigua National Bank and the Chenango County Courthouse.

Former students, including Doug Adil and Steve Zane, generously allowed the reproduction of their photographs. Michel Audy, Head of the Canadian Inventory of Historic Buildings, Parks Canada, supplied some excellent bibliographic and resource material. Many other people shared their photographs, including my wife, Susannah Falk Shopsin.

PREFACE

After years of looking for an authoritative book on historic preservation in America—one that realistically treated practical matters of restoration, extended use, and adaptive use of old buildings—I became convinced that the only way to have that book was to write it.

I have had two goals for this book. The first has been to organize, synthesize, and streamline much elusive information that architects, preservationists, and students of architecture must know and that amateur preservationists, owners of historic buildings, and save-the-neighborhood activists would find useful in working with professionals.

The second goal has been to fill a gap in the preservation bookshelf by providing a concise, jargon-free, and realistically illustrated book focused on the kinds of buildings that, more and more, architects are called on to restore, preserve, or adapt. These are not major landmarks and monuments, but ordinary old commercial buildings, houses, public buildings, and churches that are endearing elements of American townscapes and cityscapes.

Long before the terms *extended use* and *adaptive use* came into favor, and long before historic preservation became a discipline of architecture, some property owners in some American communities were practicing creative preservation. One such example is the building on the cover of this book, the Daniel Low & Company building in Salem, Massachusetts. Built in 1826, it then housed two banks on the first floor and the First Church of Salem on the second floor. Daniel Low moved his mail order jewelry and silversmith business into a small section of the first floor in 1874, and the company purchased the entire building from the First Church in 1924. Through successive changes of management in the 112 years since Daniel Low & Company moved in, the building has been respectfully preserved and adapted to contemporary use.

My hope is that readers of this book will take a closer look at the ordinary old buildings we have inherited, see beyond their grime and neglect, and discover their great potential for enhancing the contemporary environment.

The Daniel Low & Company building was constructed in 1826 as the fourth home of the First Church of Salem, Massachusetts. The thrifty congregation provided for income-producing tenants, originally two banks, on the ground floor. Daniel Low moved his fledgling company into a small section in 1874 and expanded into the former church space on the second floor when the church sold the building to him in 1924.

This old engraving, from a 1932 Daniel Low & Company catalog, shows the twin-towered facade with its High Victorian, Gothic style slate roofs and iron cresting. The adaptability of both the building and the Daniel Low firm, which introduced souvenir spoons and mail-order catalogs to America, has been rewarded with longevity. The landmark structure still dominates Essex Street, the principal commercial thoroughfare of Salem. It was converted into a mall in the 1970s.

CONTENTS

CHAPTER ONE

PRACTICING PRESERVATION

PROFESSIONALISM, PRIORITIES

Sentiment for historic preservation in America seems to be increasing as the end of the twentieth century approaches. The reclamation of major landmarks continues to be an exciting, attention-getting part of the historic preservation field. But more and more, as preservation awareness deepens in the society, it is ordinary, everyday buildings, interiors, and neighborhoods of the past that architects and preservationists are being called on to save and reuse.

Given the potential growth of professionally supervised preservation and resoration, it is paradoxical that only a small minority of practicing architects who are being approached with commissions to restore, recycle, or add to old buildings have had formal training in historic preservation. This book is a first course in preservation for the practicing architect and a first course in architecture for the active preservationist—and for everyone who cares about professionalism in preservation and restoration.

Figure 1-1 Facade restoration, Clinton Hill, Brooklyn, New York. Gentrification is flourishing in old neighborhoods all over North America. Finding qualified craftsmen and exercising patient and knowledgeable supervision are real challenges to owners and their architects.

THE PRACTICE OF PRESERVATION

A specialization in restoration and rehabilitation adds some unique concerns to the normal problems of an architectural practice. The restoration architect assumes a curatorial responsibility for structures of historic significance. Safeguarding the historic integrity of an old building undergoing restoration and recycling complicates even the most routine aspects of construction. In new construction if work is botched or unsatisfactory it can be redone. In restoration or rehab work, if the original old plaster ceilings, brick chimneys, wood paneling, parquet floors, or hardware are damaged or ruined they may be irreplaceable.

Contemporary construction laborers are often unfamiliar with traditional methods of construction and are capable of causing irreparable damage as a result (Figure 1-1). The restoration architect assumes a tremendous burden of professional responsibility in coordinating and supervising preservation work. Finding skilled architectural personnel to manage preservation projects in the office and in the field is both difficult and costly. The production of the drawings and specifications requires great care, particularly in the accuracy of field dimensions. Because old buildings are so irregular, standard-sized items from manufacturers' catalogs seldom fit and many ordinary components have to be custom fabricated off the job site. This requires a great deal of coordination to make sure that critical items arrive at the right time to keep the project moving on schedule.

Specialized Professional Training

Thus far neither architectural school accreditation groups nor state professional licensing boards have officially defined historic preservation as a recognized specialty. Many schools accept candidates with liberal arts and American history backgrounds and do not require formal architectural training as a prerequisite. However, a growing network of colleges and universities across the country has developed both undergraduate and advanced programs in historic preservation and is sending graduates out into the field. Because specialized training in historic preservation only began in the late 1960s, many of today's practicing architects have had no formal training in preservation although they are familiar with many vestigial traditional methods of construction. Plaster and lath, gravity and steam heat, and wood-framed screens and storm windows are part of their practical experience and they do not need a slide lecture or textbook to recognize these things.

It is probable that the trend toward requiring continuing professional education will provide the opportunity for mature practitioners to catch up with their newer colleagues. Since the boom in downtown revitalization and recycling has become a nationwide industry, fewer architects can afford to restrict their practice to new buildings.

Most young architects begin their careers with remodeling projects until they can secure larger commissions for major new buildings. The distinction between a remodeling and a restoration is more than just a matter of semantics. Restoration and preservation require a knowledge of and familiarity with historic styles and traditional construction techniques and detailing. It can be embarrassing when an experienced architect is taken to task by an architectural review board or landmarks panel because of an insensitive design proposal for an old building or a new one in a historic district.

Obtaining Design Approvals

For most architects, getting the commission seems perhaps the most difficult task. However, getting the commission approved is an equal challenge, leaving the actual construction in third place. Securing design approvals is far more frustrating than coping with zoning and building codes. There may be reasonable disputes over the proper interpretation of codes, but there are no secure guidelines on aesthetics. Because matters of taste and style are so subjective, the design professional is often caught in the middle between satisfying the client's program requirements and pleasing the reviewers. Often, even the most modest objectives are unattainable because of politics and community pressures.

Tactics. Television and the media have penetrated the hinterlands and stimulated citizen participation in environmental and preservation issues. In addition to the architect's conventional role as a designer, the architect is now placed in the position of media coordinator, with a greater emphasis on presentation and salesmanship. Because so much depends on winning approval, one cannot afford to lose. More time and money must be spent on professional services than is usual, since a proposal must be carried much further than the preliminary design stage just to confirm its feasibility. As a matter of tactic, it is very difficult to modify a design once it has gained approval. You and your client must be prepared to stick permanently with whatever you propose. The best strategy in advance of a public review is to prepare some alternatives or concessions for trading should the necessity arise. A cooperative attitude and a willingness to study and review the board's recommendations are preferable to a rigid and uncompromising posture.

Presentation. How elaborate should your presentation be? Do you need drawings, scale models, color renderings? There are many factors to consider even if your resources are unlimited, which is seldom the case. An extravagant presentation can have a negative effect, since it arouses suspicion that a "heavy sell" is really a cover for something that otherwise might not be acceptable. At the other extreme, an amateurish presentation may cast doubt on the ability of the applicant to undertake a project of good quality. Gauging the appropriateness of the presentation is a matter of cautious judgment and prior experience. Chapter 4, Getting Approvals, should be studied. If you are unfamiliar with the procedures of a particular reviewing agency, it might be worthwhile to attend a hearing before making your presentation. You will be more secure in understanding what is expected and better prepared as a result.

Too often, architects mistakenly assume that the general public and even the citizen appointees on review boards can read drawings. The average person has a very difficult time translating a two-dimensional set of plans into a three-dimensional reality. If the complex symbols and graphics on presentation drawings do not facilitate communications with lay audiences, verbal descriptions do not add much either. Scale models showing the proposed project in the context of its surroundings can be very helpful in explaining a project (and cheaper in the long run) than countless elegantly rendered architectural drawings. Scale figures, trees, and cars also help provide reassuring and recognizable reference points for nonprofessional reviews.

Practicing Preservation for Profit

Most architects with a professional interest in historic preservation are active in civic affairs and local preservation matters. This can lead to some professionally awkward situations where it becomes difficult to draw the line between public spiritedness and personal interest.

There is a tacit expectation that professionals are supposed to be charitable when it comes to aiding good causes. There is a corresponding perception that craftsmen, such as carpenters and plumbers, must be paid for their efforts, but that professionals should not charge for their services. This is a ticklish issue for architects and preservation professionals. Organizing a landmark rescue effort is extremely time consuming, particularly if the project involves neighborhood groups and community activists. The only thing a professional *has* to sell is advice; donated consulting time is lost time. As a matter of professional ethics, the donation of preliminary professional services is considered unfair competition, especially if it results in securing an actual architectural commission.

Because of the special nature of preservation work, a tremendous amount of time is consumed by meetings, committees, and public hearings. It is wise to offer these preliminary professional services at an hourly rate. The client may insist on establishing an upset price or a cap on professional fees, but this is still preferable to a flat fee arrangement. Until the scope of the work is defined it is difficult, if not impossible, to determine a fee based on a percentage of construction costs.

One of the major differences between the professional services required for historic preservation work and new construction is the amount of preliminary survey and investigation work necessary to determine project feasibility before the normal design process begins. Unless the initial studies on old buildings are carefully reviewed and the architect is thoroughly familiar with the existing conditions, serious problems can arise in the actual design process. And because the architect may later be held accountable for having missed some hidden flaw or weakness, it is extremely important that experienced office staff oversee the work. Premiums to architects for professional liability insurance to cover errors and omissions have risen prohibitively along with the alarming increase in claims against architects.

John Mesick, FAIA, a principal in Mendel-Mesick-Cohen-Waite, an Albany, New York, firm responsible for a large volume of preservation projects, feels that it is essential to have personnel continuity throughout the project. Mesick does not recommend the typical practice of shifting the project from a field team to a design team to a production team as it moves through different stages in the architect's office.

Finding experienced preservation-trained architects to staff an office is more difficult than hiring ordinary architecture school graduates. More qualified personnel generally earn more, so that office operating expenses will be greater for a specialized historic preservation practice. This is one good reason that a more diversified architectural practice, which also takes commissions for new construction unrelated to historic preservation, makes sense. Although experience seems to indicate that new construction is more profitable than historic preservation, during periods of economic recession when new building slows down, restoration and recycling projects seem to increase.

Contractual Arrangements

The National Historic Resources Committee (HRC) of the American Institute of Architects (AIA) has been active over the years in developing specialized contract documents for historic preservation projects. The AIA's standard reference *Architect's Handbook of Professional Practice* includes Section C-1, Preservation Practice, which has been prepared by HRC as a guide to relate the special requirements of historic preservation to standard contractual documents. Because the legal ramifications of all these specialized contractual forms have been carefully studied, it is advisable not to alter them without the advice of a lawyer.

Design/Build

Rather than waiting around for a recycling commission or trying to persuade clients on the potential merits of a neglected old building, many architects have joined with real estate professionals and investors to form design/build teams. One of the principal advantages of the design/build team is that it can respond more quickly to development opportunities than can the more conventional approach.

Arrangements vary, but typical design/build teams include an architect, a real estate lawyer, a builder or general contractor, a developer or real estate investor, an engineer, and sometimes a sales and marketing person. With this combination of expertise, it is possible to find old buildings that seem to have good potential for recycling and assess the feasibility of acquisition, probable zoning variances, code compliance requirements, and construction cost, and then estimate the cost of financing and the rate of return on the investment. Because the group is preparing the drawings and construction documents for its own use they are not as elaborate as those normally required for competitive bidding.

In a typical design/build team the architect provides professional design services as his/her contribution to the project, in return for part ownership in the project. Formerly the AIA frowned on actual participation in construction, fearing that economic considerations might conflict with sound professional judgment or influence an acceptance of inferior workmanship. Section C-1 of the *Architect's Handbook of Professional Practice* does not mention the design/build approach.

Nevertheless, the design/build concept has become more accepted and has been sanctioned by professional licensing authorities and the companies that provide professional liability insurance coverage. Before entering into a design/build team arrangement, however, an architect should seek expert advice about the impact on his/her professional liability insurance coverage. (See Summary of Topics Covered by Section C-1, Preservation Practice.)

DESIGN APPROACHES

Every old building presents the architect and owner with an array of possibilities for its further use. Although more complicated, a building located in a historic district, whether officially designated or not, presents the same range of choices: restoration, extended use, adaptive use, or thorough modernization. Working together on an individual

SUMMARY OF TOPICS COVERED BY SECTION C-1 OF THE *ARCHITECT'S HANDBOOK OF PROFESSIONAL PRACTICE*, PRESERVATION PRACTICE

1. Preservation terms and definitions.
These clarify the sometimes-fuzzy confusion over the new preservation terminology

2. Preservation services
Architectural services are divided into distinct phases, with an emphasis on the special requirements of preservation:

a. *Predesign and research*

b. *Schematic design*

c. *Design development*

d. *Construction documents*

e. *Bidding or negotiation*

f. *Construction and contract administration*

g. *Post-construction services*

h. *Final report*

3. The preservation team
A very thorough listing gives all the specialists who might be required in a complex project, with a description of their individual roles.

4. Education and training
Brief discussion includes formal education, internship, continuing education, and self-education for architects.

5. Preservation clients
Potential public sector, institutional, and private sector organizations requiring architectural preservation services are briefly listed.

6. Compensation for architectural services
Contains some helpful suggestions about the advantages of varying payment methods: lump sum, percentage of construction cost, and multiples of direct expense at different phases in the project.

7. Supplementing AIA standard forms of agreement
Special conditions and provisions to be added to the standard AIA agreements are suggested at the appropriate phases for preservation-related services, such as special reports, laboratory work, documenting, cataloging, protection.

project, the architect and owner will have to decide which is the right choice, given the owner's objectives, the structure's stylistic characteristics and structural limits, and the community's zoning and other legal constraints. Guides throughout this book will help architects and owners to assess the possibilities and choose the best role for the continued life of a building or district. The case studies in Chapter 5, some of which are referred to below, offer concrete examples of a variety of design approaches.

Restoration

There is always the possibility of meticulously restoring a building to its original condition. This can mean removing later additions, replacing lost material or parts, and making hidden repairs. A true restoration also requires that the building's original function be continued or restored. Although there is a limited need today for technologically obsolete powder mills, horse stables, and piers for ocean liners, many other building prototypes may continue to serve their original purpose.

Extended Use

Many old buildings continue to function as schools, churches, courthouses, hotels, and railroad stations although they may require some updating to extend their use (Figure 1-2). This can involve alterations and repairs—some of them major.

Despite the fact that a building may not be functionally obsolete for its original purpose, it may be abandoned because the community no longer perceives a need for it. More often, it is condemned simply for seeming old-fashioned. This was true of the Chenango County Courthouse in Norwich, New York, which is an example of successful preservation by extended use (Chapter 5).

Figure 1-2 Restored shopfronts, Empire State Building, New York, New York. Architect William C. Shopsin, in coordination with the New York City Landmarks Preservation Commission, developed master plans to serve as flexible guides to the changing requirements of several prominent commercial structures. A program of upgrading the street level shops of the Empire State Building is gradually restoring their original appearance.

Adaptive Use
The term *adaptive use* is relatively new but the concept is old. It means, quite simply, providing for a new function in an old building or, if a district is involved, many buildings. Usually extensive interior and exterior renovation are necessary (Figure 1-3). The Bank Center in Pittsburgh (Chapter 5) illustrates one approach to rescuing small downtown commercial buildings; the Seattle Garden Center Building (Chapter 5) is another example.

Extensive Modernization
Some of the most celebrated American skyscrapers are more than half a century old. The Empire State Building, the Wrigley Building in Chicago, and the Prudential Building in Buffalo all have undergone extensive modernization programs to furnish them with high-speed elevators, air conditioning, and new telephone and electrical systems (Figure 1-4). On a more horizontal scale, both Carnegie Hall and Radio City Music Hall in New York City have undergone extensive refurbishing. Many large revitalization projects require a combination of restoration, adaptive use, and infill building in order to succeed. New York City's South Street Seaport and Philadelphia's Society Hill have taken two decades to achieve the momentum of success.

Figure 1-3 Recycled former loft manufacturing building, New York, New York. Artists in New York City were originally attracted to run-down and partially vacant nineteenth-century manufacturing and warehouse structures because they offered large, cheap combined studio and living spaces. Now loft living has become trendy, and elaborate renovations have attracted the more affluent.

CRITERIA FOR CHOOSING A PRESERVATION STRATEGY

The choice of approach to preserving an old building is dictated by the economics of real estate, unless there is a special subsidy or tax incentive that alters the financial picture. Location is a key factor in determining value. The conversion of the former Greenwich Village Presbyterian Church to apartments in Portico Place (Chapter 5) is a case in point. The project was successful because of the desirability of New York City real estate. A comparable investment in an identical church in a suburban area would be harder to justify financially.

Heavy subsidies made the Chenango County Courthouse extended-use project economically feasible. But if the program demands had required more space, it might not have been possible to fit everything within the confines of the existing old building. A new addition might have adversely affected the project finances as well as created an awkward architectural design problem.

Code compliance can also be a decisive factor in project economics. The most promising adaptive use may cause tremendous additional expenses. The cost of installing sprinklers in the McKim, Mead & White mansion in the Whitefield case study (Chapter 5) almost depleted the profitability of the project. Many local building codes are tolerant of pre-existing, nonconforming uses, but are rigid on adapting to new ones.

These factors vary so greatly from one old building to another that it is impossible to give a straightforward guide to choosing a specific preservation approach for a project.

Figure 1-4 Recycled downtown club building, Salt Lake City, Utah. Vacant lots in urban areas can often be used to advantage, as in this example in Salt Lake City. New windows in a formerly blank wall face out onto a landscaped outdoor cafe on a new plaza. A skylight has been inserted in a former light court.

DESIGN OPTIONS

The first test of any design approach in preservation is whether the proposed program space requirements fit the confines of the existing old building. If it is a snug fit, it may be necessary to consider a minor addition to accommodate present and future needs. This may be more easily accomplished if the old building is freestanding. Depending on the character, materials, and detailing of the building, even a minor addition might be difficult to blend in or conceal. In the case of maintaining an extended use, the updated program needs may fit comfortably in the old building, but achieving compliance with government regulations, such as access for the disabled, may be difficult without compromising either the interior or the original exterior appearance.

Additions or Infill

Sometimes the new program requirements are so much greater than the existing old building can accommodate that the size of the needed addition will overwhelm the original structure (Figure 1-5). Establishing the acceptable limits of new infill or an addition is a very difficult and subjective decision that depends, in part, on the general aesthetic standards of the community in which the project is located. In densely built urban historic districts, abrupt changes of height and scale may be tolerated. But if the contrast is too great, adjoining property owners may protest and it may be necessary to consider other alternatives, including finding another location for the proposed project and adapting the old building to another use. In suburban or rural areas the site may permit construction of modestly scaled additional structures to accommodate the new program requirements as an alternative to tampering with the original old building (Figure 1-6).

Figure 1-5 Major addition to the Daily News Building, New York, New York. This Art Deco landmark, designed by pioneering modern architect Raymond Hood, has had several major modifications since its construction in the 1930s, among them the 1960s major addition designed by Harrison & Abramovitz to harmonize with the original materials. Recently, Skidmore, Owings and Merrill has recycled the former newspaper printing plant portions into additional office space.

Figure 1-6 Drive-in added to Mission style bank, Berkeley, California. Often, in order to remain economically viable, a commercial landmark structure must be modified. A new drive-in teller window has been sensitively designed to complement the original 1920s building in a popular regional architectual idiom.

Partial Demolition or Replacement

While the proposed program is more often too tight a fit for an old building, occasionally the old building is oversized. In many former commercial lofts and industrial buildings now being converted to residential use, as in New York City's SoHo cast-iron district, it is necessary to demolish a portion of the structure to provide adequate light and ventilation. Partial demolition may also be required if a portion of an old building is seriously deteriorated or structurally unsound. This may be the result of neglect, weather exposure, vandalism, or fire. Careful analysis may reveal that an old building is literally beyond repair or, more likely, that the costs of reconstruction far exceed those of replacement. In that case, abandoning the project is the only rational choice.

Moving an Old Building

Because a great deal of an old building's history is related to its site and surroundings, the decision to move it should not be made until all other alternatives have been exhausted. With small structures, there are often extenuating circumstances for which moving is the only realistic alternative to demolition.

Road widening, shorefront erosion, and land reclamation schemes may require only relocation on the same site. This is less problematic than removing a historic structure to another site or "restoration village," or placing fragments of it in a museum. Since the restoration of an entire community exceeds individual resources, those who remove an abandoned historic structure frequently justify the dislocation as a rescue; they fail to realize that they are ruining the future possibility of reviving a historic district. Restoration of the long-neglected eighteenth-century district of Newport, Rhode Island, was only possible because so much of its original architectural fabric was intact (Figures 1-7 and 1-8).

Modern technology permits us to commit follies in the moving of old buildings that our ancestors, who were limited to teams or horses, steam engines, and barges to shift their buildings, could not dream of. While this does make it possible to relocate even very large masonry structures, such as churches, it is still a very dubious and costly undertaking, which might be far better invested in restoring or adapting the building in situ.

PRESERVATION PRIORITIES

Everyone who is concerned with preservation continually confronts decisions and makes hard choices. Individuals involved in the preservation process have distinct points of view and particular interests and do not necessarily agree with each other. These individuals or groups include design professionals, property owners, preservationists, members of planning commissions and zoning boards, and city and town officials.

Close to the heart of every preservation project there is the question of what *can* be salvaged from what remains. And even if a building or neighborhood can be physically preserved or restored, the ultimate question has to be satisfactorily resolved among all parties at the predesign stage. The question is: Does this project make economic sense?

The chapters here prepare both professionals and laypersons to make informed critical decisions on the basic questions of whether and how to preserve and restore old buildings. The case studies in Chapter 5 show how a number of owners, preservationists, and architects have made predesign decisions and then have gone on to preserve old buildings and to convert them to new economically and socially healthy uses.

Figure 1-7 Eclectic architectual mix, Newport, Rhode Island. One of Newport's great charms is its mix of architecturally diverse structures. Ranging from the early eighteenth century to the late nineteenth century, these buildings have survived side by side with only minor modifications. Recent restoration projects have respected and maintained these later modifications.

Figure 1-8 Restoration of eighteenth-century houses, Newport, Rhode Island. In order to preserve the historic ambience of Newport's eighteenth-century neighborhoods, the Duke Foundation bought and meticulously restored the exteriors, the architectural detailing, and even the color schemes of the old houses. The restored houses are rented to carefully selected tenants who are not permitted to make any modifications.

CHAPTER TWO

LEARNING THE CONTEXTS

SITE, COMMUNITY, CLIMATE

The question of what is worth saving is a philosophical and practical one at the very core of any historic preservation project. The question must be asked over and over again throughout the often-long process of attempting to preserve or adapt an old building so that it remains a functioning part of the environment.

WHAT MAKES IT WORTH SAVING?

There is a tendency to become so focused on the rescue of a threatened old building that one loses sight of the surrounding environment (Figure 2-1.) Most older buildings, except some rural ones, do not exist in isolation from other structures. The extent to which an old building's size, materials, color, and architectural style relate comfortably with its neighbors is a key factor in the success of any preservation effort. When making any intervention in a historic setting, no matter how minor, you must constantly ask how it will fit in with the surroundings.

One of the first things to investigate is how the structure originally functioned and whether this usage is still viable. If you come to the conclusion that restoring the original function is not feasible, you must realistically analyze how the old structure relates to today's needs and possibly to the future of its community. Most old buildings are functionally obsolete long before they are physically and structurally worn out.

Quite often preservationists find themselves in the quandary of trying to rescue an architecturally and historically significant structure in a setting that is severely deteriorated. Today it is generally felt that the removal of an old building from its original site should be considered only as a last resort after all other possibilities are exhausted. Preservationists recognize that sense of place is essential and that historic buildings should not be treated as movable artifacts. Not only must the individual character of an old building be respected but also its relation to local tradition and context must be recognized.

Figure 2-1 Half a house, Bedford-Stuyvesant, Brooklyn, New York. As a practical matter, half a house may be better than none, but the loss is difficult to come to terms with aesthetically. Without a symmetrical scheme the design options for replacement are extremely limited.

Recognizing the elements that give an area its special character depends on careful, sensitive observation. Some places are harmonious in their uniformity, others dramatic in their contrasts. In areas or places that have been inhabited for centuries, there is a pervasive sense of local or vernacular building traditions that have developed slowly over time. The dominance of a particular indigenous building material is immediately apparent. Roof silhouettes—gabled, mansard, or flat—reflect local climate conditions as well as stylistic preferences. A predominant scale is also an important ingredient in the visual cohesiveness of a particular area. Subtle gradations of small to large in the open countryside will be perceived just as dramatically as gigantic towers in crowded cities.

Regionalism is still a strong and resilient force in contemporary American society. It is diversity and quirkiness that we most cherish in old buildings, which are one-of-a-kind survivors in an age of mass production. Each project is different and each must be evaluated in survivial contexts.

EXPLORING LOCAL ARCHITECTURAL CHARACTER AND TRADITION

On your first site visit you can make an informal windshield survey of the locale of the project. A quick tour will allow you to observe the predominant character and range of architectural idioms in the area. You should visit not only the property that is the subject of a proposed project, but the surrounding neighborhood and community as well. (See Figure 2-2.) It may take several visits to become familiar with local architectural character and tradition. In some areas guidebooks and local histories may be of some assistance and provide you with a portable reference; if such references are not available, it is wise to take along a camera to record distinctive structures in the vicinity. (See Checklist 2-1.)

Respecting Local History

In most areas that have been inhabited for a long time, recognizable vernacular building traditions have developed despite modifications and additions over the years. Take note of the most common building materials. Typically either wood or masonry predominate, usually either painted wood clapboard or plain brick. In certain regions, indigenous materials stand out: logs, wooden shingles, fieldstone, stucco, or mud-plaster adobe.

Figure 2-2 Old stone houses, New Paltz, New York. The old village center of New Paltz contains a cluster of seventeenth-century houses constructed of local fieldstone by Huguenot settlers. One of their unique features is the steeply pitched medieval gabled roof containing several attic storage levels.

Checklist 2-1
DEVELOPING A VISUAL PROFILE OF A BUILDING

As part of a windshield survey, a visual profile should be developed by describing the building as if seen through a zoom lens, beginning with the distant view and moving item by item to the smaller details. Below are some examples of what might be included.

Relationship to Landscape
☐ on a hilltop ☐ on a hillside ☐ in a valley
☐ on level ground ☐ on rolling ground
☐ hidden ☐ partially obscured ☐ in an open space

Rooflines and Silhouettes
☐ pitched roof ☐ flat roof ☐ dormers ☐ cupolas ☐ gables
☐ towers ☐ chimneys ☐ silos ☐ vents (tall or low)

Size and Form
☐ large ☐ small
☐ spread out ☐ compact massing ☐ clustered

Materials
☐ wood ☐ stone ☐ brick ☐ stucco ☐ other ( )
☐ textured ☐ smooth
☐ dark ☐ light
☐ bright against landscape ☐ blends into background
☐ organic use of materials

Openings and Details
☐ flush doors ☐ recessed doors ☐ glazed doors
☐ small windows ☐ large windows
☐ decorated windows ☐ plain windows
☐ porches ☐ columns ☐ balconies ☐ entablatures ☐ pediments

Most of what is now being preserved in the United States was constructed during the late eighteenth, nineteenth, and early twentieth centuries. Throughout the seventeenth and eighteenth centuries, most communities tended to limit themselves to the use of locally available materials and local craftsmen who were guided by imported pattern books and carpenters' manuals. During the early colonial period the importing of raw materials or ready-made architectural elements was not only slow and undependable but also prohibitively expensive. (See Figure 2-3.) As a result, only in the construction of major public buildings and residences for the wealthiest individuals, such as Southern planters and New England merchants, would imported materials and architectural elements be incorporated.

By the early nineteenth century the American economy was more firmly established and less dependent on imports. Some mechanization and standardization of building components had been achieved. It was no longer necessary to have all building components custom-made, which meant that resulting building materials were more available to everyone. This, along with the growth of road and rail networks across the country, created new distribution patterns. Until regional economies developed their own manufacturies, cast-iron decorative elements were sent from New York City to New Orleans (see Figure 2-4), and whole prefabricated wooden or cast-iron buildings were shipped around Cape Horn to San Francisco during the Gold Rush.

Figure 2-3 Mansion, Salem, Massachusetts. The residences of the prosperous eighteenth-century traders were designed and furnished in the latest English fashion.

Figure 2-4 Antebellum cast-iron balustrade, Savannah, Georgia. Distinctive ornamental and decorative architectural elements were often imported when they were not locally available.

Whereas the process of trying to identify the style of a seventeenth- or eighteenth-century building is limited to recognized prototypes, researching nineteenth-century American architecture is much more complex than that of earlier periods because of its diverse origins and eclectic sources of inspiration. Mass printing of illustrated periodicals and books resulted in a broader circulation of new ideas, styles, and popular trends. The old established elite were no longer the sole arbiters of fashion and popular taste yielded greater diversity and inventiveness. Thus it is not surprising to discover enormous Eastern-style mansions in Portland, Oregon; Galveston, Texas; or in remote mining towns in Colorado (see Figure 2-5). Quite often these somewhat exotic transplants bear no relation to the local vernacular building tradition in matters of style or choice of materials.

Figure 2-5 Queen Anne style mansion, Salt Lake City, Utah. A wealthy Salt Lake City couple admired a mansion on New York City's Riverside Drive and commissioned its architects, Lamb and Rich, to build a copy. The original has long since been demolished, but the copy has been adapted to law offices and a photographer's studio.

Recognizing Styles

In addition to cataloging the distinctive characteristics of local buildings, it is useful to accurately classify them according to recognized architectural styles. To the architectural historian, buildings can be *read* by their distinctive details, proportions, and choice of materials. Most historic buildings fit given stylistic categories, despite regional variations and idiosyncracies. For example, Federal and Greek Revival styles flourished from Maine to Alabama. In all areas, classically inspired structures contain similar architectural elements such as columns, porticoes, carved wreaths, Greek key motifs, and cornices. There are a number of good guides to American architectural styles (see Bibliography).

Identifying Visual Concerns and Sense of Place

When considering the visual effect that a proposed project may have on an existing historic environment, there are major differences between town and country. A small lot in an urban neighborhood located within a rigid grid of streets offers few options in placement and the spaces between buildings may be little more than an alley. By contrast, a village or suburban setting where freestanding structures are separated by greenery, fences, yards, and lawns is far less restricted and permits objectional aspects of a project to be concealed. Indeed, in rural areas structures may exist in total isolation or small clusters related more to the landscape than to property lines or roads.

One of the most perplexing challenges to preservationists is restoring and sustaining large, isolated white elephants that have suffered long periods of neglect and remain as impractical relics of an extravagant era. Sometimes it is only an individual property, but it is a much more serious problem when an entire commmunity is out of fashion or in economic decline (Figure 2-6). Rhinebeck in Dutchess County, New York; the great camps of the Adirondacks; and some New Jersey shore resorts contain or are made up of great hulks of dilapidated mansions. In such places, the prognosis for successfully rescuing one individual property may be entirely dependent on the fate of the surrounding community.

EVALUATING THE COMMUNITY

When exploring a community, it will be quite apparent from the current physical condition of ordinary buildings, landscaping, fences, walls, and paving if a neighborhood or district is affluent, stable, or declining. For example, the presence of abandoned mills, factories, railroad stations, or vacant stores usually indicates a weak local or regional economy. The surroundings must be taken into account when planning for preservation, because it is unrealistic to expect that a single project is going to revitalize a whole neighborhood. (See Checklist 2-2.)

Figure 2-6 Rustic gatehouse, Craggsmore, New York. The abandoned former estate of the Hudson River School painter George Inness has now fallen into the state of picturesque ruin.

Checklist 2-2
ASSESSING A BUILDING'S RELATIONSHIP TO COMMUNITY

- ☐ Does the proposed project follow the predominant character and range of architectural idioms in the area?
- ☐ Does the particular building have architectural elements and distinctive details, such as columns, porticoes, or cornices, that identify it as belonging to a nonregional style category?
- ☐ Is the scale of the proposed project compatible with the surrounding structures?
- ☐ Is the intended use of the building consistent with local zoning laws?
- ☐ Does the surrounding community show signs of economic decline or instability, and will it be able to support the project?
- ☐ Are the building's materials and mode of construction compatible with the area's climatic and environmental conditions?
- ☐ Does the site present problems in subsoil drainage?
- ☐ Does the building's proximity to property lines or heavily traveled roads impose limitations on the proposed project?

Figure 2-7 Pike Place Market, Seattle, Washington. The colorful Pike Place Market, a multitiered structure poised on a steep hillside facing the Puget Sound, almost succumbed to an urban renewal project. The restored and enlarged Art Deco style Seattle Garden Center building is featured in Chapter 5.

Urban Communities

Countless downtowns, disrupted by post-World War II urban renewal and exodus to the suburbs, have scarred cities large and small. Scattered old buildings in vast urban wastelands are almost accidental survivors of massive demolition programs. By the late 1970s, many communities fortunately recognized the failure of the bulldozer approach to city planning and were actively encouraging the revival of center-city residential and commercial neighborhoods. An effort was made to incorporate historic remnants in large-scale redevelopment projects. Major waterfront restoration projects have created new downtown tourist centers: New York City's South Street Seaport, Boston's Faneuil Hall Market, Philadelphia's Society Hill, Baltimore's Harbor Place, Savannah's Factor's Walk, San Francisco's Ghiradelli Square, and Seattle's Pioneer Square and Pike Place Market (Figure 2-7).

While restoration has revitalized many formerly elegant urban neighborhoods and created fashionable "new" historic districts, the trend to gentrification has been accompanied by dislocation of the poor. Several cities have sought to mitigate the social impact of gentrification through urban homesteading programs in which abandoned and neglected residential properties are offered to low-income families in exchange for so-called sweat equity. However, the transformation of former commercial and manufacturing loft structures into middle-class housing, shops, galleries, and office space continues to result in the loss of many already scarce unskilled jobs in many urban areas.

Complexities of Urban Restoration. The successful restoration of urban areas is a very complex process requiring the coordination of many professional disciplines. Creating harmony between the existing old structures and major new construction requires that a designer take into account many considerations beyond the normal requirements of an individual project. Urban structures abut each other, frequently

Figure 2-8 New York Public Library and Knox Building. The Knox Building, by architect John Duncan (also Grant's Tomb), frames the landscape terraces of the massive Beaux-Arts New York Public Library by architects Carrere and Hastings.

Figure 2-9 Republic National Bank, New York, New York, Attia & Perkins architects. The bank's new tower wraps around the old Knox Building, dwarfing it and eliminating its mansard-roofed silhouette, as well as the vista down Fifth Avenue to the Empire State Building. The relationship between big and small is often more significant than the contrast between new and old.

sharing a common wall or separated by only an alleyway. Abrupt contrast between neighboring properties cannot be concealed; dissimilar color, texture, size of openings, and ornament are immediately apparent (Figures 2-8 and 2-9).

Of all the design factors to be considered in urban settings, contrast of scale and disparity between high and low are probably the most difficult to resolve. The economics of real estate development seem always to push for bigger buildings. In the more homogeneous settings found within urban historic districts, downzoning to maintain uniformity of height is generally more acceptable. It may, in fact, be impossible to maintain a harmonious relationship between an isolated small individual historic structure surrounded by new construction.

For example, Brooklyn, popularly known as the "Borough of Churches," still has neighborhoods dominated by domes, steeples, and spires. Across the river in Manhattan, even massive St. Patrick's Cathedral is dwarfed by the towering office and apartment buildings that dominate New York City's skyline (Figure 2-10). Whereas the topography of the land dominates the countryside, producing variations in height, the city is an almost totally artificial environment in which even the terrain and planting is manipulated and chosen.

Figure 2-10 St. Paul's Chapel, New York, New York. Once the tallest and most prominent structure of New York's prerevolutionary era, St. Paul's Chapel is now dwarfed by surrounding towering office buildings. There is no way to provide harmony with such a tremendous contrast of scale, but the church's identity is preserved because it remains freestanding in its own graveyard.

Suburban/Countryside Communities

The process of suburbanization has slowly eroded the charm and picturesqueness of once rural countrysides, small villages, and tiny hamlets. As vehicular traffic increases along winding country roads, dignified old houses, churches, and schools become stranded among shopping centers, used-car lots, motels, and fast food drive-ins.

Even the most carefully planned individual preservation effort cannot ensure that a restoration will not be lost in a sea of visual pollution. Through careful review and revision of local zoning ordinances, it is possible to achieve guidelines for preserving local scenic routes and design control of roadscapes. Unfortunately, most small towns do not have the resources or the local expertise to cope with the pressures of development and so are unable to legislate and enforce the kinds of ordinances required to control what happens along major routes that pass through their communities.

Special Problems of Old Mill Communities

America was so vast in the earliest period of settlement that little thought was given to the placement of buildings. Until the Industrial Revolution created the great mill complexes, most local water-powered sawmills and wind-powered gristmills were so modest in scale that they fit picturesquely into the countryside. As long as the cottage industries persisted, little thought was given to the separation of living and working. In many areas utopian communities, such as Old Economy and Hopewell Furnace in Pennsylvania, thrived on the combination.

At first, mills and factories were located near waterways because of abundant energy and ease of transport. The steam engine and the locomotive created greater mobility, allowing industry to move inland. The post-Civil War industrial boom spread mechanization and industrialization across the country. The primary fuel of the nineteenth century was coal, which left its gritty traces on rural manufacturing communities and great cities alike. In the twentieth century the switch to oil and gas has further polluted the atmosphere. The introduction of the assembly line and automation have had even more drastic and dislocating consequences.

One of the difficulties in finding new uses for abandoned factories and mills is their location in proximity to residential areas, often in conflict with current zoning ordinances. Even though little thought was originally given to the separation of living and working, subsequent zoning regulations have rigidly enforced these distinctions in most localities. Particularly in urban areas, the conversion of former loft manufacturing structures to art galleries, studios, and apartments is frustrated by protracted delays and entanglements with contradictory zoning and building code regulations. Before planning an adaptive use that is obviously quite a departure from the existing one, it is important to be familiar with local zoning and building codes. These issues are discussed in greater detail in Chapters 3 and 4. Recycling and reuse of former manufacturing and commercial structures is not solely an urban problem, since many abandoned mill sites may be in rural or suburban communities. The individual solutions, however, vary greatly depending on location and local economic and social factors that may extend far beyond the conventional parameters of the architect's area of expertise.

Figure 2-11 Log tavern, Bedford, Pennsylvania. The heavily forested Allegheinies in western Pennsylvania provided the early settlers on the pioneering route to the West with ample building materials.

IMPLICATIONS OF CLIMATE AND LOCAL SOIL CONDITIONS

Recognizing the impact of local climate and soil conditions is another significant part of planning successful adaptations of historic structures. Throughout the United States tremendous extremes of climate, terrain, and vegetation have led to distinctive regional architecture. Many of the architectural responses to these environmental factors have been shaped by the ethnic origins of the builders. Much American building is a recall of past foreign traditions—Spanish adobe, Swedish log cabin, and Dutch brickwork, to suggest a few (see Figure 2-11).

However, from the 1830s until the 1930s a predilection for historic revival styles and eclecticism dictated much of our architecture. This led to some awkward and impractical imports—such as Mediterranean stucco villas (see Figure 2-12) in wintry New England locales; tiny-windowed, wood clapboard Cape Cod cottages in subtropical Florida; and elegant Louis XV French limestone pavilions, à la Versailles, perched on the salty, windswept sand dunes of the Hamptons on Long Island. Many of these exotic experiments have not weathered well and require constant maintenance or extensive replacement with more indigenous materials. Poor judgment in site placement has resulted in tragic losses of life and property. Along the Eastern seaboard, shore-front erosion is a constant threat; in the West, earthquakes and mud slides cause havoc. Elsewhere, poor subsoil drainage conditions cause flooding, uneven foundation settlement, and structural damage that is often impractical if not prohibitive to remedy. There are circumstances in which we must admit that, although intellectually and aesthetically appealing, an old building has not stood the test of time and should not be preserved.

Environmental pollution, grime, and acid rain have caused extensive damage to brownstone, glazed terra cotta, copper, and other traditional materials and construction assemblies. Ironically the introduction of such energy-saving measures as insulation, storm windows, and tightly sealed interiors has accelerated the deterioration of many older buildings (see Chapters 7 and 9).

Figure 2-12 Caramoor, Katonah, New York. North of New York City a rambling estate was inspired by Venetian, Gothic, and Mediterranean models, created as a background for a connoisseur's eclectic collections of European art and furnishings, to provide an intimate setting for chamber music. Together with its Italian gardens it has been preserved as a house museum and performing arts center.

FOCUSING ON THE PROJECT AND ITS SITE

Once you have explored the region and the immediate surroundings, you can identify most of the external factors influencing a property. Predominant local character and vernacular building traditions, the use of indigenous materials, neighboring structures, siting and relationship to the landscape, and the impact of climate can all be taken into account in planning for preservation.

Next, assess the project site to determine what limitations the placement of existing old buildings and their proximity to property lines impose on the project. For example, one of the most difficult problems in preserving old farmhouses and agricultural complexes is that they are usually located too close to a road. Beautifully surrounded by mature trees and plantings, what was once considered a pleasant country lane or dirt road has become a narrow paved road with hairpin turns, where the bucolic has been supplanted by gasoline fumes, nighttime headlights glaring in the windows, and shrieking brakes. All of these make the property far less attractive for residential occupancy. If the site is large enough, it may be worth moving the main residence further back from the road.

Finally, make a detailed examination of the old building itself to determine the internal factors and physical limitations that might restrict its restoration, upgrading, and adaptation to contemporary functional requirements. These considerations include the method of construction, physical condition of the structure, size, exisiting layout, and ceiling heights.

Once these determinations have been made, you will have to study your proposed program requirements and choose one of several options:

1. Restore or adapt the building without any significant additions to its original volume.

2. Add space to the existing building in order to satisfy your program requirements.

3. Partially demolish and replace portions of the existing building.

4. Move the building on the site or to another site because there is no viable alternative to maintaining it at its present location.

CHAPTER THREE

ASSEMBLING THE RESOURCES

SURVEYS, DATA, PROFESSIONAL EXPERTS

To avoid wasted time and money, your first order of business is to find out how much information is already available about your building and its site. The most important information is at the site, that is, in the physical condition of the structure itself. Before anything else you must thoroughly investigate and become familiar with the old building. Chapter 6, Evaluating the Structure, deals with the specifics of physical inspection. If there is any doubt about a building's structural soundness, it may be necessary to engage a structural engineer whose report should be the basis of your most fundamental decisions. Once you have ascertained the physical condition of the structure, you can begin to investigate information about its original appearance and develop a program for its restoration. If a historic resource survey, which is an inventory of historically and architecturally significant structures, exists for your project area, you have a head start.

LOOKING FOR INFORMATION ON THE ORIGINAL BUILDING

Because making a survey can be tedious and frustratingly slow, take every opportunity to investigate local resources and historic societies, and seek out knowledgeable longtime residents of an area to determine if any surveys or records exist. Unfortunately, most historic structures are not accompanied with old plans, photographs, detailed descriptions of original appearance, color schemes, and so on. Even when such information exists it is often incomplete, inaccurate, and amateurishly organized.

If local and regional sources yield no results, move upward through state and federal agencies (parks and recreation departments) that have been mandated to conduct historic preservation surveys, often as part of the environmental review process of land-use regulations. In many regions ancient card indexes are being transferred to computerized database systems. (See the Resources section for a list of agencies.)

There is no absolute standard for surveys. They run the gamut from a typewritten list of historically significant structures to elaborately photographed, mapped, and recorded studies that include scale models and renderings of proposed restorations of individual buildings or entire historic districts.

The more information that is available to the designer the easier it is to successfully relate the individual project to its surroundings. If the size, materials, rhythm of openings, signs, street graphics, and lighting are already well documented in the survey, then the design process can move ahead much more rapidly with less effort wasted on finding out what is desired and acceptable. (See Checklist 3-1).

Checklist 3-1
LOCATING SURVEYS AND DATA

HISTORIC RESOURCE SURVEYS

- An inventory of historic or significant structures within an area
- Initiated at local level (city or county) and assisted by state and federal agencies
- Valuable for overall data on resources in district or region
- Aids in comparative research of similar buildings, structures, and sites
- Elucidates styles, periods, and construction methods

Local sources:
Historic, civic, or professional organizations

State sources:
State Historic Preservation offices; statewide preservation organizations

National sources:
HABS/HAER
Prints and Photographs Division
Library of Congress
Washington, D.C. 20540
(provides originals or copies of art and documents)
HABS/HAER
National Park Service
U.S. Department of the Interior
Washington, D.C. 20013
(provides information on existing surveys)

OTHER HISTORIC DATA (WRITTEN AND PICTORIAL)

- Provides guides to accurate restoration or replacement of missing or altered elements of old buildings
- Includes documents—wills, deeds, tax records, construction and inspection records; maps and prints; measured drawings, floor plans, renderings, blueprints; photographs

Local sources:
Municipal offices; architectural and photo archives in libraries, museums, universities, colleges, historic societies, and other specialized organizations; local newspaper files; family albums; illustrated local and regional histories and guidebooks

State sources:
State Historic Preservation offices; statewide preservation societies and professional organizations

National sources:
COPAR (Cooperative Preservation of Architectural Records)
Prints and Photographs Division
Library of Congress
Washington, D.C. 20540
(directs researchers to appropriate archives)

Library of Congress
Prints and Photographs Division
Washington, D.C. 20540
(contains thousands of prints, negatives, and transparencies documenting rural and urban architecture from 1800s to 1940s; reproductions may be obtained from Library's Photoduplication Service)

Library of Congress
Geography and Map Division
Washington, D.C. 20540
(contains maps and illustrated atlases of nineteenth/ and twentieth-century cities and counties)

National Register of Historic Places
National Park Service
U.S. Department of the Interior
Washington, D.C. 20013
(maintains records of structures and districts with HRHP-designated status)

National Trust for Historic Preservation
1785 Massachusetts Avenue, N.W.
Washington, D.C. 20036
(maintains a library with research facilities)

ARCHAEOLOGICAL RECORDS

- Valuable for historic interpretation of sites and construction methods

Local sources:
Private developers; civic groups

State sources:
State archaeologist; archaeology department of state university

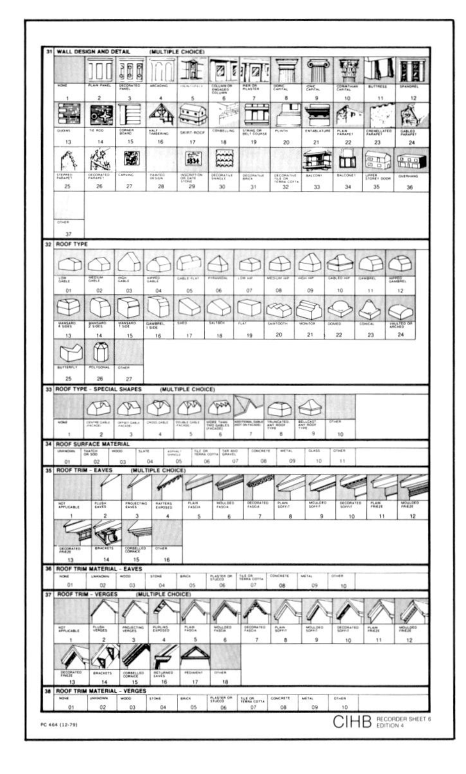

31 WALL DESIGN AND DETAIL (MULTIPLE CHOICE)

32 ROOF TYPE

33 ROOF TYPE - SPECIAL SHAPES (MULTIPLE CHOICE)

34 ROOF SURFACE MATERIAL

35 ROOF TRIM - EAVES (MULTIPLE CHOICE)

36 ROOF TRIM MATERIAL - EAVES

37 ROOF TRIM - VERGES (MULTIPLE CHOICE)

38 ROOF TRIM MATERIAL - VERGES

PC 464 (12-79) CIHB RECORDER SHEET 6 EDITION 4

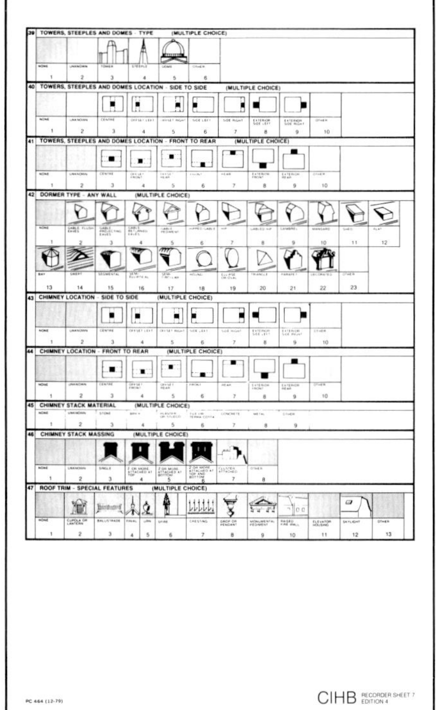

39 TOWERS, STEEPLES AND DOMES - TYPE (MULTIPLE CHOICE)

40 TOWERS, STEEPLES AND DOMES LOCATION - SIDE TO SIDE (MULTIPLE CHOICE)

41 TOWERS, STEEPLES AND DOMES LOCATION - FRONT TO REAR (MULTIPLE CHOICE)

42 DORMER TYPE - ANY WALL (MULTIPLE CHOICE)

43 CHIMNEY LOCATION - SIDE TO SIDE (MULTIPLE CHOICE)

44 CHIMNEY LOCATION - FRONT TO REAR (MULTIPLE CHOICE)

45 CHIMNEY STACK MATERIAL (MULTIPLE CHOICE)

46 CHIMNEY STACK MASSING (MULTIPLE CHOICE)

47 ROOF TRIM - SPECIAL FEATURES (MULTIPLE CHOICE)

PC 464 (12-79) CIHB RECORDER SHEET 7 EDITION 4

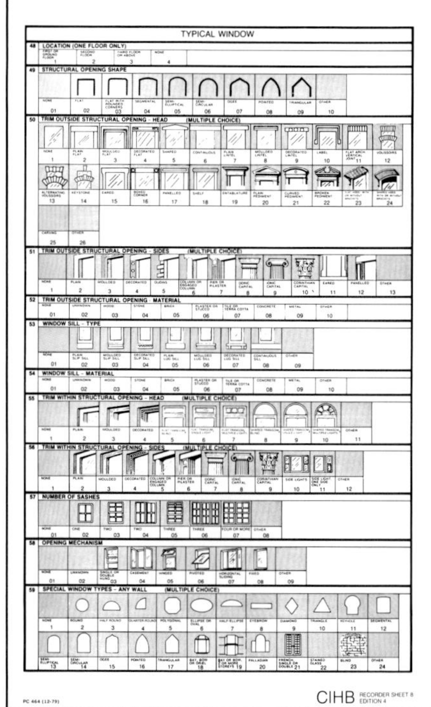

TYPICAL WINDOW

48 LOCATION (ONE FLOOR ONLY)

49 STRUCTURAL OPENING SHAPE

50 TRIM OUTSIDE STRUCTURAL OPENING - HEAD (MULTIPLE CHOICE)

51 TRIM OUTSIDE STRUCTURAL OPENING - SIDES (MULTIPLE CHOICE)

52 TRIM OUTSIDE STRUCTURAL OPENING - MATERIAL

53 WINDOW SILL - TYPE

54 WINDOW SILL - MATERIAL

55 TRIM WITHIN STRUCTURAL OPENING - HEAD (MULTIPLE CHOICE)

56 TRIM WITHIN STRUCTURAL OPENING - SIDES (MULTIPLE CHOICE)

57 NUMBER OF SASHES

58 OPENING MECHANISM

59 SPECIAL WINDOW TYPES - ANY WALL (MULTIPLE CHOICE)

PC 464 (12-79) CIHB RECORDER SHEET 8 EDITION 4

Figure 3-1 Canadian Inventory of Historic Buildings, Parks Canada. This standard survey form was devised by the Canadian Inventory of Historic Buildings. To be most helpful a survey should be designed with graphic examples of easily recognizable vernacular features.

Historic Resource Surveys

Most surveying efforts are accomplished through a combination of professionals, students, and dedicated local preservationists who pool their talents to visit, photograph, record, and research the historic and architectural background of a selected area. Depending on the focus of the effort and the degree of professionalism involved, historic resource surveys vary greatly in the amount of data gathered and accuracy of interpretation. Typically, surveys are conducted to identify individual structures of historic and architectural merit as well as concentrations of significant old buildings as a means of determining the boundaries of a proposed historic district. Once the survey has been completed, guidelines and zoning ordinances can be formulated that are tailored to the unique character of the district.

Until recently the data generated by a typical historic resource survey filled file cabinets full of unwieldy reporting forms and index cards. (See Figure 3-1.) Many of these surveys are now being transferred to a computerized format that simplifies research and retrieval. Most local historic resource surveys are organized under the supervision of a state's parks agency, which maintains the actual files for liaison with the National Park Service for listing properties on the National Register of Historic Places.

A historic resource survey may not provide a great deal of specific detail or background information on individual buildings unless they are of exceptional historic or architectural interest. Even though your project is included within the survey boundaries, it may not have been studied thoroughly enough to provide all the necessary information you require. If your project is not included in the survey, contact the participants in the original historic resource survey, because it is possible that more extensive research and documentation exists that was not incorporated in the published survey.

Locating Plans and Other Data

The principal benefit of locating old plans, descriptions, or photographs is to serve as a guide to the accurate restoration or replacement of missing or obviously altered elements of an old building. In the absence of actual documentation you must depend on an educated guess as to what the original builders intended.

Depending on the age of a building, finding surviving copies of the original plans is a hit-or-miss proposition. If the building was built before the turn of the century, it is very unlikely that the original plans have survived. Even the most responsible institutional property owners seldom can produce a complete set of documents. Most cities and towns did not begin to keep organized files of building construction until the turn of the century, and fires, damp basements, moving, and periodic house cleanings have diminished the usefulness of most of these early municipal archives.

By the end of the nineteenth century, photographers avidly recorded every facet of American life. Ordinary buildings were seldom documented but they are often glimpsed in the background of family portraits and local events that were the photographer's actual objective. Newspapers, historic societies, and family albums can produce valuable images. Once you have thoroughly searched these sources and are satisfied that all archives, both private and public, local and regional have been exhausted, move on to the state level and contact the State Historic Preservation Officer (SHPO), who usually is the director of the parks and recreation agency of the state. The SHPO's offices are generally repositories of valuable information and are well connected to a network of knowledgeable architectural historians, archaeologists, and preservationists in the state.

Figure 3-2 HABS measured drawing. Greek Revival doorway, Greenwich Village historic district, New York, New York. This type of precise archival documentation is important not only for recording the details of architectural elements in case of loss, but also as a prototype for reproducing missing pieces in other surviving buildings.

After you have exhausted the resources at the local, regional, and state levels, move on to the federal agencies. Prior to the involvement of the state governments in archaeological and historic preservation issues, most of the initiatives were at the federal level. During the Great Depression one of the projects established by the Works Progress Administration (WPA) was the Historic American Buildings Survey (HABS). Unemployed architects were assigned to teams all over the country to survey, measure, and record distinctive early American architecture. (See Figure 3-2.) The principal focus of this vast documentation project was the colonial period through the early nineteenth century.

A similar federal project, the Historic American Engineering Record (HAER), was initiated to document exceptional examples of engineering and industrial design, such as mills, bridges, dams, canals, railroads, piers, factories, and the like. The archives of both HABS and HAER are maintained in the Library of Congress in Washington, D.C. (copies are available at a nominal charge). This recording effort continues to the present but at a considerably reduced rate because of the escalated costs of sending qualified teams out into the field. Modern methods, including photogrammetry, have permitted more extensive documentation with less field and drafting-room time.

The restoration of historic public buildings and parks all over the country was also undertaken as part of the Civilian Conservation Corps (CCC) and the National Recovery Act (NRA) to aid the unemployed during the Depression. Two notable examples of these public works efforts, now designated historic districts, are popular tourist attractions: the picturesque former flood-control ditch, Paseo del Rio in San Antonio, Texas; and Cliff Walk in Newport, Rhode Island, which allows a stunning view of the harbor as well as a close glimpse of the great mansions. Although most of the records were sent to Washington, D.C., sometimes a thorough search of local archives will yield detailed field notes and records of NRA- and CCC-funded projects that may prove useful in planning restoration efforts.

Archaeological Records

Sometimes, when all other kinds of investigation and historic research have been fruitless, there may be no choice but to dig. A professional archaeological dig can be a major undertaking, particularly if the site is urban. The earth beneath several skyscrapers in the historic core of New York City's financial district has yielded fragments of seventeenth-century ships as well as hordes of other artifacts.

The tedious process of digging, sifting, and carefully documenting these sites is very costly and time consuming. For the extremely valuable Lower Manhattan sites, the costs of the investigations were largely underwritten by the developers as a trade-off for various zoning concessions. In suburban and rural areas it can be very difficult to find the financial resources needed to carry out a proper archaeological investigation.

The archives of archaeological research are much more difficult to locate than those for historic buildings. Often investigations of individual sites have been made for academic or historic interest and may not be publicly recorded. However, sites of archaeological, historic, and architectural interest will often be identified by government agencies because they lie in the path of proposed public works improvements such as highways, dams, utility distribution systems, sewer and waste disposal systems, and land reclamation projects. Federal law now mandates that Environmental Impact Statements (EIS) must be prepared for any public works project receiving a subsidy from the federal government. The EIS is reviewed and approved by both state and federal environmental agencies to determine that public funds will not be used for any public works project that would have negative impact on a significant archaeological site or historic structure. If a proposed project is found to have an adverse effect it must be reviewed by the President's Advisory Council on Historic Preservation, a federal agency that may recommend mitigating measures. In order to avoid potential conflicts and protracted delays, many state and local agencies have initiated their own historic resource studies or inventories before determining the location of proposed public works projects. Both early buildings and sites and recognized landmarks of modern architecture, such as Miami's Art Deco district, have been identified in this way.

ASCERTAINING LEGAL STATUS

Another basic consideration of any potential project is its current status with local, state, and federal jurisdiction. All but the most rural locations now have some form of zoning and land-use regulations. Old buildings usually predate the ordinances and do not always conform to current regulations. Most ordinances permit continued usage and minor alterations to a pre-existing nonconforming structure but severely restrict any major enlargement of the building or expansion of its usage.

The degree of zoning nonconformity caused by an undesirable usage may be compounded by some aspect of an old building's construction that violates modern life-safety codes. Usually fire rating codes and requirements are the principal cause of difficulty. If substantial changes are planned, some form of variance, or exemption, will have to be sought and granted in order for an old building to achieve code compliance.

Landmarks legislation is another determinant of the project's status. It is primarily local ordinances that vary greatly in scope, enforcement, and penalties. A project may comply with zoning and building code regulations and still be deemed incompatible by a landmarks or architectural review commission. Findings based on aesthetic considerations and matters of taste are highly subjective and require a spirit of patience and compromise. A historic preservation consultant can be of great help here.

DETERMINING TAX STATUS

The tax status of old buildings is of serious concern to property owners and developers. Assessments in older, run-down neighborhoods may be low, but they will not remain that way once recycling and gentrification get underway. Many local and federal programs have provided tax incentive and abatement programs to stimulate neighborhood revival. These programs are always subject to modification and termination. It is wise to thoroughly check the status and applicability of these programs early in the planning process so that you can design the project to maximum economic benefit.

ASSEMBLING THE PROFESSIONAL TEAM

Although much of the initial project research can be done by dedicated volunteers, which holds down costs, there is no substitute for what experienced professionals can provide. Assembling the professional team to study, analyze, and design the proposed project is perhaps the most critical factor in the success of your project. The composition of the professional team will vary greatly, depending on the project's character, location, and perhaps most of all, its budget. Typically, members of the professional team with a specialized background include the following: historic architect, architectural historian/researcher, archaeologist, architectural photographer, surveyor, and historic engineer. (See Checklist 3-2.)

The Historic Architect

Professional training in historic preservation has existed only since 1965, when a graduate program at Columbia University was established. Many mature practicing architects have had no specialized education to prepare them for preservation work. Therefore, it may be advisable for an architect undertaking a preservation-oriented project for the first time to seek the assistance of an architect who is a specialist or consultant in preservation.

Demonstrated experience and familiarity with preservation are essential. In addition to studying published works and brochures it is a good idea to arrange visits to completed preservation projects and also to meet and talk with clients directly. Letters of reference are no substitute for personal contact. Most local chapters of the American Institute of Architects (AIA) will provide a list of members with demonstrated experience in historic preservation.

The Architectural Historian/Researcher

Local historians and teachers can be extremely knowledgeable and helpful in locating archival material. Finding someone with expertise in architectural history is often more difficult. Many architectural historians have moved beyond their former academic boundaries to become more directly involved in all phases of historic preservation, including research, design, and construction. Their background and education and familiarity with architectural history and styles can relieve the architect of much time-consuming and tedious research. The architectural historian's testimony may also be of great value in appearances before landmarks commissions and architectural review boards.

The Society of Architectural Historians (SAH) has organized chapters in most major urban areas and at universities with architectural and art history faculties. SAH maintains an extensive network of specialists all over the country and can make referrals appropriate to the specific requirements of your project.

The Archaeologist

In most preservation projects of major historic significance, the search for historic information and documentary evidence on which design decisions will be based can be very frustrating. With very early colonial sites, the only evidence often lies within the structure itself or buried beneath it. The archaeologist can solve many of these mysteries and provide useful guidance.

Many archaeologists have specialized training and areas of interest such as cultural anthropology. Finding an archaeologist with a specific interest in and familiarity with historic American architecture can be difficult. Many state parks and recreation agencies can provide some archaeological assistance. Most archaeologists are involved in academic research and their limited time available for fieldwork is usually committed far in advance.

Archaeology is a patient process that cannot be rushed or maneuvered to suit construction timetables. This may cause scheduling difficulties, since archaeology is governed by weather conditions even more than are construction and restoration. Also, construction should be

Checklist 3-2
ASSEMBLING THE PROFESSIONAL TEAM

Project Evaluation	Maintaining the integrity of a historically significant structure complicates even routine aspects of restoration or recycling. It may be necessary for an architect who is attempting a preservation-oriented project for the first time to use the services of a **historic architect** on a consultant basis.
Historical Research	Even if a historic resource survey exists, additional research may be required for a specific project. The architect may require the services of an **architectural historian** to assist in locating existing archives.
Archaeological Investigation	For sites of major historic significance, a **historic archaeologist** may be needed to excavate or examine evidence at the site. An archaeologist can find hidden structures, evaluate and date the evidence, and help validate any necessary documentation.
Site Survey	If no site survey of the boundaries exists, or if the survey is so old that it does not accurately represent the current status of the property, it may be necessary to hire a **surveyor**. A topographic survey taken at 2 foot intervals and indicating the location of major trees may also be needed in order to proceed with detailed planning, particularly if the site is not relatively level.
Measured Drawings and Photographic Documentation	If the project is complex, it may be necessary to hire outside technical experts to assist in preparing measured drawings and photographic records. An **architectural photographer** can provide rectified photographs; **experts in photogrammetry** can provide measured drawings using special equipment; **experts in computer graphics** can provide enhancement of old photographic evidence.
Structural Evaluation	To ensure that concealed structural and mechanical systems do not present insurmountable problems to rehabilitation or adaptation, it may be necessary to let a preliminary construction contract involving a **structural engineer** and/or a **mechanical engineer**. In addition, it may be necessary to hire, on a contractual basis, a **historic engineer**, someone specialized in traditional construction methods to advise on the restoration of old structures and the inclusion of new mechanical systems.

phased so that it does not disturb areas of the site that are critical to the archaeologist's investigations, such as the locations of early wells, privies, and waste dumps.

Figure 3-3 Architectural photographer at work. Architectural photographer Steve Zane reflected in a mirrored fireplace overmantel in one of the Villard Houses in New York City. The documentation, photographic and measured recording of this turn of the century McKim, Mead & White landmark is one of the most elaborate projects following HABS standards since the Great Depression.

The Architectural Photographer

Historic preservation requires specialized techniques of recording photography. Advances in photography have meant that much of what once had to be painstakingly measured and drawn by hand can now be documented instantaneously and with far greater accuracy. Photogrammetry, which eliminates visual distortion, is indispensable in measuring inaccessible parts of buildings that formerly required scaffolding to photograph. By means of darkroom techniques, photogrammetric images can be transferred directly to the architect's reproducible drawings and annotated for use in construction (Figure 3-3).

All of this has resulted in tremendous savings in time and labor as well as providing simpler, more practical construction documents for field supervision of restoration work. Finding an architectural photographer is sometimes difficult because the specialized equipment for photogrammetry is elaborate and expensive. In many cases rectified photography, which uses more conventional equipment, is sufficient.

Even if measured drawings exist, a thorough photographic record of conditions prior to restoration is important in case of loss or damage. Because the tracing or transferring of the photogrammetric images is the most costly part of the process, often only the photo negatives on special archival-quality film are produced and stored for future reference.

Another specialized aspect of restoration-related photography is the reproduction and analysis of historic photographs. Often a faded old photographic print may be the only evidence of a long-vanished architectural detail, interior, or landscaping. Experimental techniques, including computer graphics, are being used to enhance these images and generate reproductions that are superior to the original prints.

The Historic Engineer

In engineering today, just as in architecture, there is increasing specialization. In some locales it may be quite difficult to find a civil engineer who has practical experience in the restoration of traditional construction methods such as cast-iron and mill-timber framing or structural terra cotta. Restoring, enlarging, replacing, or moving old buildings require a particular expertise and you should be prepared to look long and hard for the right person. It is even more difficult to find a mechanical engineer who is sensitive to the needs of restoration. Accommodating existing conditions of old buildings and minimizing the intrusion of new equipment require a great deal of patience and ingenuity, because stock solutions taken right out of a catalog are seldom satisfactory. It may be necessary to seek out specialized historic engineering expertise on a consulting basis.

Organizations of Experts

If your contacts fail, there are several places to turn for recommendations. Local and state arts councils have contacts with knowledgeable individuals, professionals, and groups who can assist in research and documentation resource surveys or an archaeological dig. The National Trust for Historic Preservation (NTHP), as well as state and local preservation groups, offer technical assistance and lists of qualified consultants. Many colleges and professional schools now offer advanced training in historic preservation, conservation, and archaeology taught by specialists who are available as independent consultants. Professional organizations, such as AIA and the Association for Preservation Technology (APT), are additional sources of referrals (some of these organizations are listed at the end of this book).

PRELIMINARY COSTS

It is important for the architect to make the client aware that restoring and adapting old buildings is more complex than new construction in both the design development stage and in the actual construction process. Many more preliminary professional services are required to document existing conditions. Sometimes it is necessary to get a preliminary construction contract in order to establish the condition of concealed mechanical and structural systems.

One of the pitfalls of working with old buildings is that unanticipated, concealed defects regularly surface in predesign studies. These either restrict the adaptability of a proposed use or add considerably to the anticipated cost of the project. Since most owners sell property "as is" and generally will not permit a prospective buyer to do structural investigations, purchasing an old building is a calculated risk. Depending on the willingness of local officials to give a preliminary review of zoning and building code compliance, there remains a considerable risk that your proposed adaptive use may not be granted a required variance. The conclusion of all the preliminary investigations may show that the proposed project is technically impractical or economically infeasible, in which case the investment in these preparatory services is lost.

Figure 3-4 SoHo historic district, New York, New York. SoHo was the first commercial/manufacturing district in the country to be designated as a historic district. Gradually, the manufacturing that at first coexisted with the influx of artists, galleries, and restaurants was driven out. The building, shown here, pictured in 1978, is now entirely residential above the ground floor.

PRESERVATION ALTERNATIVES

Ultimately, it is money that controls the fate of an old building. Assuming it gets approval (which is the subject of Chapter 4), a building can continue in the same use or can be adapted to a new use.

Extended Use

Many old buildings, particularly institutional ones, continue to serve their original purpose. Frequently institutional owners recognize that their old church, school, or government buildings are no longer functioning efficiently and are tempted by conventional expedience to reduce maintenance and operating costs by replacing their old structures with modern ones. More enlightened institutional owners now invest in design studies to find out if they can rehabilitate their facilities.

Because of their symbolic and historic significance, many institutional structures are truly irreplaceable and must not be sacrificed. However, some functional and program requirements may have to be tailored to the limitations imposed by the old buildings.

New Use

To generate the funds necessary to restore a building, a change of use—new use or adaptive use—may be the only way to increase its economic viability. In commercial historic districts such as Savannah, Georgia's waterfront and New York City's SoHo (see Figure 3-4), the original warehousing and manufacturing occupants have gradually been supplanted by restaurants, shops, and residential tenants.

The design study will tell whether an old building can be extended or enlarged without seriously compromising its original appearance and character. The relationship to its existing surroundings and context must also be taken into account in this determination. Obviously, whether the proposed enlargement makes good economic sense is critical to the decision. This is a patient process of trial and error until a successful balance is achieved. Unfortunately, circumstances do not always permit adequate time to work out an ideal solution.

Figure 3-5 Abandoned warehouse, Troy, New York. One of a group of severely deteriorated and fire damaged nineteenth-century riverfront warehouses originally slated for demolition as part of a major downtown urban renewal project. This structure and its neighbors got a last minute reprieve from a developer attracted by a federal tax incentive program.

Last-Minute Rescue

Frequently time runs out and a desperate last-minute effort is made to rescue a landmark structure threatened with demolition. To operate under these pressures is far from ideal from a professional standpoint. Because of the time shortage, and usually money too, the preliminary research and investigation cannot be accomplished and many important decisions must be made on the basis of superficial observations and assumptions. Hasty judgments can lead to errors that are extremely expensive to rectify once the crisis is over and the normal pace of project design development is under way. If there are significant old buildings in your community, do not wait until the marketplace determines their fate (Figure 3-5).

CHAPTER FOUR

GETTING APPROVALS

REGULATIONS, ECONOMIC AND SOCIAL ISSUES

Most places, from tiny rural hamlets to the most densely populated urban centers, have some form of land-use ordinances. Everywhere, zoning laws are increasingly restrictive and complex to administer—density, bulk, height, setbacks, and special features such as signage, plazas, pools, and parking are only some of the regulations.

Getting a project approved can be a straightforward process if the proposal conforms to the regulations. However, growing public concern over environmental pollution, visual blight, and overcrowding has led to stringent laws and review procedures. You can be caught in the web of overlapping and sometimes contradictory regulations of local, state, and federal legislation. Growing support for historic preservation has added yet another layer to the approval process.

PRINCIPAL ELEMENTS IN THE APPROVAL PROCESS

The first step in any project is to secure a copy of the latest local zoning and building code requirements. In addition to reading the ordinances it is important to become familiar with the prevailing interpretation and day-to-day implementation of the written regulations. Simply stated, "as of right" is what an ordinance permits you to do without special review. Since most old buildings predate the modern ordinances, they are often classified as nonconforming in some aspect of their construction, size, site placement, or use. Any project which is not "as of right" introduces uncertainty, the need for special government review, and the possibility of delay. These factors may result in additional costs for consultants—not only for design but also for lawyers.

It is therefore essential to quickly determine what aspect of a project is not "as of right" and to investigate what, if any, practical alternatives exist. If these issues cannot be resolved it may be necessary to abandon the project.

The following hypothetical situations give some of the most common difficulties of obtaining approvals (actual case examples can be found in Chapter 5).

1. The existing use is no longer permitted by the current zoning. A pre-existing nonconforming use may be tolerated, but any substantial remodeling or enlargement may not be allowed. *Example:* A mill or a warehouse is located in a residential area. The building may be restored but cannot be expanded for extended use.

2. A pre-existing use may be "as of right," but the construction of an old building does not conform to the current building code requirements. *Example:* An old mansion is to be used as an inn or guest house, but it lacks sufficient exits and is of nonfireproof wood construction. The building must be updated for new use.

3. An old building conforms to existing usage and building code standards but does not meet the current site coverage, setback, or parking requirements. *Example:* Unless an alternative provision for the required parking can be found the adaptive reuse potential of an otherwise attractive landmark may be severely restricted.

4. An old building and its site conform to local zoning and building code provisions but cannot meet the standards of a special ordinance. *Example:* An old estate with narrow winding roads and an avenue of mature trees is to be converted to condominium units and cannot comply with a special ordinance that mandates the width of access roads for emergency vehicles without destroying the historic landscape.

5. A landmark structure in an urban area is underbuilt, which means that the current zoning "as of right" allows a much larger structure on the same site. *Example:* This is a very common occurrence in major American cities today, where nonprofit groups like churches, libraries, museums, and concert halls are tempted by substantial offers from developers to demolish, rebuild, or greatly enlarge their existing landmark structures. Even if the proposed project is eventually approved, dealing with public outcry, media coverage, and court proceedings are time consuming and very costly to all parties.

6. Many localities have designated historic districts. Inevitably, in drawing the boundaries, some structures of little architectural distinction (as well as vacant lots) have been included within the district. For the owner or potential developer of such sites, as well as for other historic district residents, the concept of change may become a major issue of proportions and a source of endless public debate.

In order to develop a method of dealing with these real-world situations, preservationists must grapple with the three principal components of the approval process.

- Legal and Legislative Requirements
- Economic Issues
- Social Issues

LEGAL AND LEGISLATIVE REQUIREMENTS

Since most legislation is written in general terms, its application to specific situations is always subject to some degree of interpretation. Zoning and building codes are examples of legislation that is constantly being amended and revised to achieve higher standards and embrace new concerns. Old buildings cannot always conform to the rigorous standards formulated for new construction. The legal and legislative approval process is somewhat akin to an obstacle course in which the participants must anticipate and negotiate many hurdles in order to make it to the finish line. The principal challenges in the approval process in sequence are: (1) zoning regulations, (2) building (or life-safety) codes, and (3) special ordinances and review processes.

Zoning Regulations

Since the beginning of the century, the courts have accepted the doctrine that zoning, or constraining individual property rights, is in the public interest. Recent landmarks legislation is an extension of the same concept. Zoning was originally considered a matter of public safety, separating industrial and commercial uses from residential ones. Rigid zoning also produced a logical distribution of municipal services such as public schools, parks, and mass-transit systems.

The recent trend to loft conversions of former manufacturing and commercial structures as living quarters, art galleries, restaurants, and boutiques has blurred many of these earlier planning assumptions (Figure 4-1). The gentrification of the formerly neglected nineteenth-century warehouses and cast-iron structures has accelerated the decline of these districts as manufacturing and shipping areas and created a new demand for neighborhood services, such as food markets and laundries, and municipal services, such as hospitals, public schools, day-care centers, and recreational facilities, which were not a part of these districts before. From Portland, Maine, to Louisville, Kentucky, new urban residential neighborhoods have blossomed in rundown commercial areas.

In order to permit the transformation of former commercial and manufacturing structures to new uses, the old ordinances that rigidly distinguished the two must be modified. A flexible approach to mixed-use zoning is essential, but it cannot be applied by granting variances on a piecemeal basis. A uniform policy based on compatibility with existing occupants is the key to a successful zoning strategy during the transition period.

Zoning and land-use legislation is enacted at the prerogative of local governments, and most ordinances have special provisions for circumstances in which full compliance with the regulations would constitute a hardship. You should approach the review process with a cooperative attitude and a willingness to make some compromises. The review and approval procedures generally require public hearings. These receive local media coverage. A professional presentation can often be the decisive factor in securing an approval, and good public relations coverage is important in extremely complex or controversial situations.

If an application is denied, it may be appealed at the county or regional level only if it can be proven that local review commissions acted in an arbitrary or capricious manner in the administration or interpretation of their ordinances. Make every effort to avoid this kind of confrontation, since it usually involves court actions and can cause considerable delays. If you succeed in gaining an approval by appeal, it can generate resentment and cause considerable difficulties in dealing with local officials during the actual construction.

Figure 4-1 SoHo, New York, New York. The gentrification of SoHo, a former manufacturing and warehouse district, represents a unique ambiance for a historic district. Previously picturesque or bohemian historic districts such as New York's Greenwich Village or New Orlean's Vieux Carre were old residential neighborhoods containing small buildings. The vast interior spaces of the loft buildings found in SoHo permit a much greater variety of adaptive uses.

Building Codes

The essential purpose of building, or life-safety, codes is to guarantee some degree of conformity to basic engineering standards for sound construction, ventilation, sanitation, and fire protection. Codes are constantly evolving as new materials are developed. There is also protectionist lobbying that influences codes and industry standards. Insurance underwriters, organized construction labor, building products manufacturers, and other special interest groups exert strong pressures on new code legislation. Building codes that deal with life-safety requirements are very specific. Often regulated at the level of state government, these are enforced by local building departments with on-site inspections.

Because there is always a tendency to upgrade and modernize code-safety requirements, it becomes more and more difficult to achieve compliance in old buildings. In certain areas of the country, special review procedures have been developed to cope with alternatives for historic buildings. But despite years of efforts toward creating a so-called uniform code, none has been universally adopted. As a result of widely different code requirements, restoration costs and scheduling demands can vary enormously.

Cost and Scheduling. The cost implications of code compliance are perhaps the most significant considerations in the planning process for preservation. To complicate things, code compliance is not merely a matter of the way a building is constructed but also of how it is used. Usually, less rigorous application of the modern code provisions are made for continued use, or extended use, of an old building. However, adapting to a new use may occasion a total re-examination for compliance with the most up-to-date requirements.

A proposed conversion to public use, such as for a restaurant, performing arts space, or day-care center, will focus scrutiny on fire protection, floor-load capacity, ventilation, quantity and proper location of emergency exists, adequate stair widths, and other safety requirements that can be extremely expensive (Figure 4-2). If large numbers of people are to be accommodated on an upper floor, the safety requirements become even more stringent. It is therefore essential to take into account the inherent limitations of an old building before seriously considering a radical change in its use. Many uses permitted by the zoning regulations may prove to be impractical because of difficulties with building code compliance.

Costs affect schedules, and because code compliance has such serious implications for overall project feasibility, you should try to secure some preliminary determinations of cost as early as possible in the design development process. In some communities it is possible to secure a preliminary opinion from the local building inspector. At the very least, you will be able to determine if a variance or special consideration is required. If it becomes apparent that there is no way that your project is going to get approved, it is better to be spared the costs of a useless set of architectural drawings and specifications. Once the prop-

Figure 4-2 Hotel lounge, the Helmsley Palace, Villard Houses, New York, New York. An existing secret door, disguised as a marble niche, has been transformed into a legal exit from an elaborate drawing room. A new convector has been concealed behind a *faux* marble metal enclosure beneath the window

erty has been acquired, all delays are costly. Realistically anticipate the time needed for reviewing, restudying, and negotiating that is always an integral part of the approval process.

Aesthetic and Environmental Ordinances

Most building codes steer clear of aesthetic and environmental considerations and stay with matters of structural safety, construction methods, fire protection, exit requirements, adequate lighting, ventilation, and so on. Whereas building code review is a routine bureaucratic procedure, many special ordinances require public hearings. Special ordinances are usually local laws designed to regulate some specific aspect of design or zoning, such as signage; architectural or aesthetic review; scenic, environmental, or historic preservation (Figure 4-3). There is great variety in this type of legislation, depending on local and regional precedents and patterns of enforcement. Sometimes special ordinances are the responsibility of independently appointed commissions or boards, but more often they are administered by existing agencies such as planning boards.

Prior to making any presentation before a commission or review board become familiar not only with the ordinance but, perhaps more importantly, with how it is interpreted and applied. It is also a good idea to attend a meeting before the project is scheduled for a hearing just to observe the procedure and presentation an applicant is expected to prepare. It may also be helpful to research and visit recently approved projects that are similar to your own.

Review boards can be callously unsympathetic to claims of personal or economic hardship if the project requires a variance or sets some unique local precedent (see Figure 4-4). If you are well prepared you may be able to make some public gesture or contribution to mitigate the aspects of your proposal that are perceived as negative. Even if local officials and members of the review board are sympathetic to your proposal, the community at large may not be. Providing a landscaping feature or a contribution to some local civic association may garner public support and smooth the way for approval. Don't drag out the bargaining process; in the long run, delays will cost more than the concessions you may be forced to make.

Figure 4-3 Row houses, Greenwich Village, New York, New York. As the popular concept of landmarks has shifted from the freestanding architectural monument to restoration and preservation of historic neighborhoods and districts there is a greater concern with filling in missing gaps.

Architectural Review Boards and Landmarks Commissions. Architectural review boards and landmarks commissions, whose authority is an extension of the local zoning powers, deal with visual and aesthetic matters. Many of their design decisions are subjective, making it difficult to establish standards or guidelines. Most architectural and landmarks bodies accept the premise that what was done in the past is their guide to the future. Approving the replacement or duplication of minor architectural elements such as missing doors, windows, and cornices is a simple routine. However, difficulties begin with the introduction of new storefronts, major additions, or extra stories that are without much precedent.

If there is no convenient historic precedent for the guidance of the landmarks commission, the architect has to make a much more elaborate presentation to justify the design. The architect should try to enlist the support of local civic and preservation groups prior to the public hearings to bolster his or her position. Typically, review boards reflect the more conservative rather than the liberal viewpoint of the communities that have appointed them. An architect proposing a contemporary addition to a traditional structure should usually anticipate a negative reaction.

It is helpful if a successful similar project can be located elsewhere and illustrated or, better still, visited. Most of the lay members of review panels have difficulty translating plans, renderings, and even scale models into a three-dimensional concept. It is also difficult for nonprofessionals to distinguish between matters of style and taste, particularly with some of the architectural and decorative excesses of the late nine-

Figure 4-4 New town house, Greenwich Village, New York, New York. Although architects Hardy, Holzman, Pfeiffer's design for a new town house seems to have mellowed comfortably into its historic setting, it caused a storm of controversy when it was first presented to the New York City Landmarks Preservation Commission.

teenth century. The information provided by an architectural historian can often be helpful not only in the research and preparation of documentation supporting the proposed preservation, but also in setting an authoritative tone at the public hearing.

If you and your team have done your homework well before the public hearings, you will have some indications of what objections may be raised. This may involve some redesign before the public review. By all means try to avoid a public confrontation over aesthetic matters. In your presentation, make a clear distinction between those design elements that derive from functional requirements and those that represent aesthetic preferences. Because lawyers tend to be uncomfortable about the vagueness of aesthetic review procedures, the opposition will often attempt to shift the emphasis to environmental issues as a means of delaying or defeating a project.

Many communities have enacted landmarks preservation ordinances as an overlay on existing zoning maps without taking into account the inequities that result. Clear distinctions should also be made in the application of zoning ordinances to individual landmarks and those grouped in historic districts. Since they usually are not, the burden of proof is put on the property owner. This lack of coordination can cause confusion and delay when proposing plans for enlargement or modifications that normally would be acceptable but are now rejected because the building has been designated a landmark structure.

Environmental Quality Standards. Environmental quality standards are often developed on the federal level and enforced by state and regional governments. With all these overlapping jurisdictions it is not surprising that the field of preservation law is a burgeoning area of professional specialization. Special ordinances, such as preservation tax-incentive programs, have attracted accountants and financial specialists to guide the uninitiated through the inevitable red tape. *Preservation News*, published by the National Trust for Historic Preservation, contains advertisements and a classified directory of professionals, consultants, and companies offering services.

Increasingly, environmental impact studies require considerable expertise, not only in their preparation but also in their review and interpretation. If not locally available, such expertise may be time consuming and costly to obtain. Even if you know that your project has no negative impact, demonstrating this to your critics and convincing government agencies may not be a simple matter. Many of these well-intended special ordinances are used as delaying tactics by critics and frustrated community groups who demand the preparation of elaborate environmental impact statements. In these instances it is best to consult a law firm experienced in environmental issues.

Environmental standards, specifically those aimed at increased energy efficiency, may be totally at odds with the objectives of historic preservation ordinances. Adding insulation, solar panels, or storm windows or replacing original fenestration may seriously compromise the original appearance of an old building. (Refer to Chapter 7 for more on insulation and other materials.)

ECONOMIC ISSUES

Architect Louis Kahn has said, "True economy is a beautiful thing; a budget is arbitrary." Many historic preservation enthusiasts get carried away and mistakenly assume that great old buildings are exempt from the normal economics of real estate and the marketplace. It's easy to run up the bill doing it the "right" way. Unfortunately, the same rules apply to old and new and must be respected. Your efforts may result in disappointment and considerable financial loss.

Real Estate Values

In order to establish how much investment is prudent, it is essential to be aware of comparable projects. Make a thorough investigation of existing local facilities and what they cost to build, operate, and maintain prior to making any commitment to a project. Even nonprofit institutions must operate within their budgets and establish priorities.

One of the advantages of using older buildings is the possibility of doing work in phases, rather than all at once as with new construction. However, you must determine whether the elaborateness of the proposed restoration or the additional work dictated by code compliance will generate excessive costs. It may in fact be more economical to find another building more suited to your purposes than to force an unsuitable adaptation. The case studies in Chapter 5 document a number of typical situations and discuss the compromises that were made.

Figure 4-5 Winter fire destroys Brooklyn landmark. After years of neglect the renovation of the turn-of-the-century Hotel Margaret into luxury apartments overlooking New York harbor from Brooklyn Heights, was almost complete when it was totally destroyed by a major fire. The developers' problems were compounded because current zoning restrictions would not permit a new building of similar height or bulk.

Tax Incentives

In order to offset demolition and replacement of older structures and neighborhoods, innovative government tax-incentive programs have been created to attract private investment to recycling and preservation projects. Some local governments offer partial abatement of real estate taxes or low-interest loan programs. Since 1976 the federal government has offered tax incentives through various Internal Revenue Service programs. These incentives range from accelerated depreciation to investment tax credits and have attracted private investors seeking tax-shelter benefits. Participation in these programs does have some drawbacks, since it imposes additional government reviews, approvals, and red tape to the normal approval process.

In periods of fluctuating interest rates the delays encountered in securing the government approvals are often more costly than the benefits of the incentives themselves. Careful analysis may indicate that the time saved justifies the extra costs of an unsubsidized project, and because it is free of government scrutiny and restrictions, the project is usually more appealing to lending institutions and private investors.

Insurance

Securing adequate insurance coverage is especially important in dealing with old buildings (Figure 4-5). Vacant buildings are extremely vulnerable to vandalism. It is important to seek expert advice from an insurance firm experienced in the special problems of old buildings. In working out a program of insurance, consider not only the construction process requirements but also the design and material selection alternatives that can save premiums in the future. A long-term projection may provide the economic justification for a larger initial investment in fire protection. Building codes usually set a minimum standard and satisfying these minimum requirements does not necessarily result in the lowest premiums.

Figure 4-6 Harborplace, Baltimore, Maryland. The Rouse Company has led the way in the revival of downtown and waterfront marketing and recreation areas. Whereas some of its other projects—Faneuil Hall in Boston, Society Hill in Philadelphia, and the South Street Seaport in New York—contain a core of restored and recycled buildings, Harborplace is almost entirely new construction. It has, however, stimulated the restoration of neighboring historic districts and individual downtown landmarks.

SOCIAL ISSUES

Certainly the most thorny side of the approval process is dealing with your critics. Most dissatisfaction and skepticism stems from fear of change and a desire to leave things as they are. Most communities no longer feel that new or bigger is necessarily better. But individuals are afraid that even if a proposed project is a genuine improvement, they will be displaced or somehow shortchanged. Poor communications are often the source of rumor, misunderstandings, and innuendo, which can, in turn, provoke opposition. Careful and thorough preparation involves a great deal of advance educational effort and politic public relations to lay the groundwork before the actual public hearings.

Displacement

Successful landmarks and historic preservation projects are now widely evident across the country. Large commercial waterfront restorations in San Francisco, Boston, New York and Baltimore have captured the interest of a broad cross-section of the general public (Figure 4-6). The preservation movement has made great strides in reaching beyond the elitist concerns of house museums and garden clubs. But the specter of gentrification is so overwhelming in urban areas that any restoration project is perceived as synonymous with removal of the poor and replacement with the affluent. As a result, some preservation efforts have produced a backlash that is just as strong as that which followed the bulldozer in the urban renewal era.

Community groups have been quick to recognize their ability to influence the planning and approval process. Politicians and local officials inevitably are drawn into these controversies and the architect is required to spend considerable time and diplomacy preparing, presenting, and revising proposed schemes at committee meetings and public hearings. Sometimes these situations become so complicated that a public relations firm is needed to assist in developing the presentations and securing favorable media coverage.

SUMMARY

To get approval is, in fact, to get a series of approvals, beginning with a schematic idea and resulting in a fully approved set of plans and a permit to begin construction. Although the basic sequence is the same, depending on the project and its location, the actual steps and number of approvals required vary. Some approvals are conditional and require successive reviews. Approvals at some stages are easier to obtain than others so that it is difficult to know the exact time that will be required.

The first phase of the approval process is to find out whether your proposed usage conforms to the current zoning. If not, and you plan to seek a variance first do some local research to calculate your chances of succeeding. It is better to reconsider your alternatives before submitting an application that you feel has a strong chance of being rejected. Because professional consulting and legal fees can mount up quickly, it is better to keep your submissions as schematic as possible. If you are contemplating acquiring a property competitively or at auction some of these inquiries will have to be made discreetly in advance of the bidding. Often you can get an option to allow time for a feasibility study without actually purchasing the property.

Even though it is desirable to hold down your investment in professional services during the preliminary approval phases, you must do a feasibility study in sufficient detail to obtain rough estimates of costs. The feasibility study is an ongoing process in which you begin to consider a wide variety of alternatives for your project and gradually narrow them down to a few. The feasibility study tests possibilities against the limitations imposed by the programmatic needs, zoning and building codes, construction costs, financing, and other special considerations.

Obviously the circumstances of each project are so different that it is impossible to present a formula for undertaking a feasibility study. One project may require a marketing consultant, another a landscape architect, and a third a museum curator. Particularly in the preliminary stages, when there is no organized funding plan, it may be difficult to pay for outside consultants. Nonetheless, getting the right advice from the start is critical. (See Checklist 4-1.)

There is a big difference between finding out what you can get approved and what will be the most feasible scheme to follow. Interpreting the results of the information gathered in a feasibility study should help you make these decisions:

1. What is physically possible?
2. What is legally permissible?
3. What is economically realistic?

Once you have the answers, the project should move forward with a minimum of surprises.

Checklist 4-1
EXAMINING THE LEGAL, ECONOMIC, AND SOCIAL ISSUES

LEGAL ISSUES

Zoning Regulations

- ☐ Does the intended use—whether original, extended, or adaptive—comply with current zoning regulations?
- ☐ If not, can a variance be easily obtained from a local review board, or will it be necessary to make costly appeals to county and regional authorities?

Building Codes

- ☐ Does the structure comply with current codes for structural safety, ventilation, sanitation, and fire protection?
- ☐ If not, can the structure effectively be updated to comply with modern fire and safety codes?
- ☐ Do the inherent limitations of the structure (such as floor-load capacity or provisions for egress) present obstacles to an intended adaptive use, such as conversion from private occupancy to public assembly?

Special Ordinances

- ☐ Do the local ordinances regulate other aspects of design, such as historic or environmental preservation?
- ☐ Will extensive remodeling or the introduction of major additions require documentation of historic precedent?
- ☐ Will the structure's designation as a landmark or its location in a historic district complicate plans for its enlargement or modification?
- ☐ Will the project require an environmental impact study to satisfy anticipated criticism?
- ☐ Will environmental standards, such as those aimed at energy efficiency, conflict with the objectives of historic preservation?

ECONOMIC ISSUES

- ☐ Will compliance with legal requirements result in extra costs or in scheduling delays that make restoration or adaptaton an unsound investment?
- ☐ Is the intended land use economically feasible according to current real estate values?
- ☐ Do studies of comparable projects show that the cost of the proposed project is economically justifiable?
- ☐ Do tax regulations provide incentives in terms of abatements and low-interest loans?
- ☐ Are the potential benefits of tax incentives worth the time and effort required to obtain additional approvals?
- ☐ Can better insurance coverage be obtained by using above-minimum standards for construction and materials?

SOCIAL ISSUES

- ☐ Will the project be complicated by displacing current residents or by other community concerns?
- ☐ Will it be necessary to hire a public relations firm to secure favorable reaction and influence the planning board?
- ☐ Can conflict with local community groups be avoided by providing public amenities to compensate for what are perceived as negative aspects of the project?

CHAPTER FIVE

LEARNING FROM EXAMPLE

CASE STUDIES OF PRESERVATION

The inventory of historic preservation projects in the United States has burgeoned since the 1970s. Professional architecture journals and magazines, which once emphasized only new buildings, now devote regular coverage to recycling projects.

CHOICE OF CASE STUDIES

Because the variety and distribution of successful restoration projects is staggering, only a selected sampling of restoration prototypes can be presented as case studies here. For purposes of comparison, the projects are clustered in five types: (1) churches, (2) mansions on large estates, (3) large houses on small lots, (4) commercial downtowns, and (5) public buildings.

The projects illustrate the innovative and creative process by which their architects recognized the potential of neglected old buildings and restored them for extended use or adapted them to new uses. Working within the complex web of practical limitations, the projects strikingly demonstrate the possibilities of what can be achieved. No matter how similar the building prototype may seem, the circumstances of each situation is unique and it may be impossible to successfully or economically replicate these projects.

An important lesson to draw from these case studies is the importance of securing the cooperation of the local community and its regulatory agencies such as zoning boards. Frequently, the old buildings whose charm and eccentric character we seek to preserve do not conform to modern standards. Knock-it-down-and-start-all-over-again is a much easier approach. As will be noted in the various case studies, it takes a great deal of patience, persuasiveness, and persistence to prevail in these matters.

After World War II, shifts of population away from older urban neighborhoods led to the decline and abandonment of many religious properties. These architecturally distinctive buildings, often the focal structures in a neighborhood, are threatened with demolition. Sensitive adaptation permits these venerable structures to retain their dignified landmark presence, while providing them with a new lease on life, assuring them proper maintenance, and placing them on the tax rolls.

PORTICO PLACE

PROJECT
Portico Place
Condominium Apartments
Greenwich Village historic district, New York City

ARCHITECT
Stephen B. Jacobs & Associates, P.C.

DEVELOPER
Choral Associates

STRUCTURAL ENGINEER
Alvin Fisher and
Robert Redlein, P.C.

MECHANICAL ENGINEER
Todler/Schwartz

FIFTEEN APARTMENTS NESTLE IN FORMER HOUSE OF WORSHIP

Existing Structure

An 1846 Greek Revival church building protected by the Landmarks Preservation Commission stood empty in an upper-middle-income neighborhood in New York City's Greenwich Village Historic District. Previously the Presbyterian congregation had shared its usage with a synagogue and a theater group. However, membership in the church dwindled, and efforts to find another user for the building was fruitless. Eventually the building was sold to a real estate development group.

Restoration Program

The developer determined that the most desirable adaptation was to convert the former church into multifamily housing despite the restrictions imposed by the Landmarks Preservation Commission. The commission refused permission for additional height, new dormers, or changes to the classic portico of its street facade. (The jurisdiction of the New York City Landmarks Law is limited to the exterior of religious structures.)

The conversion divided the church's former main sanctuary and attic into four floors, which together with the original basement provided 15 apartment units, some of which are duplexes and studios.

Project Finances

Compared with similar-sized apartments in the neighborhood, prices for the units were reasonable and sales were brisk. The developer made no use of tax incentives even though the Economic Recovery Tax Act of 1981 would have qualified the project for both a 25 percent Investment Tax Credit and an accelerated depreciation deduction over the next 15 years.

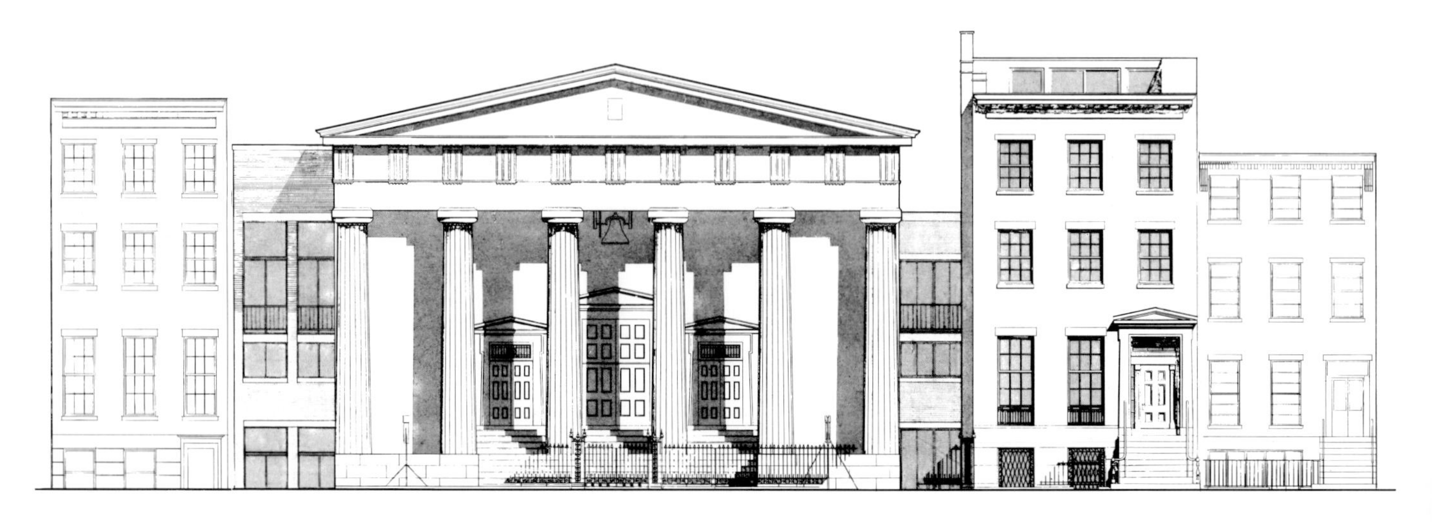

Principal elevation. This elegant drawing shows the relationship of the church to neighboring structures. The town house at the right is the former manse, which was developed into apartments while the controversy over the recycling of the church dragged on.

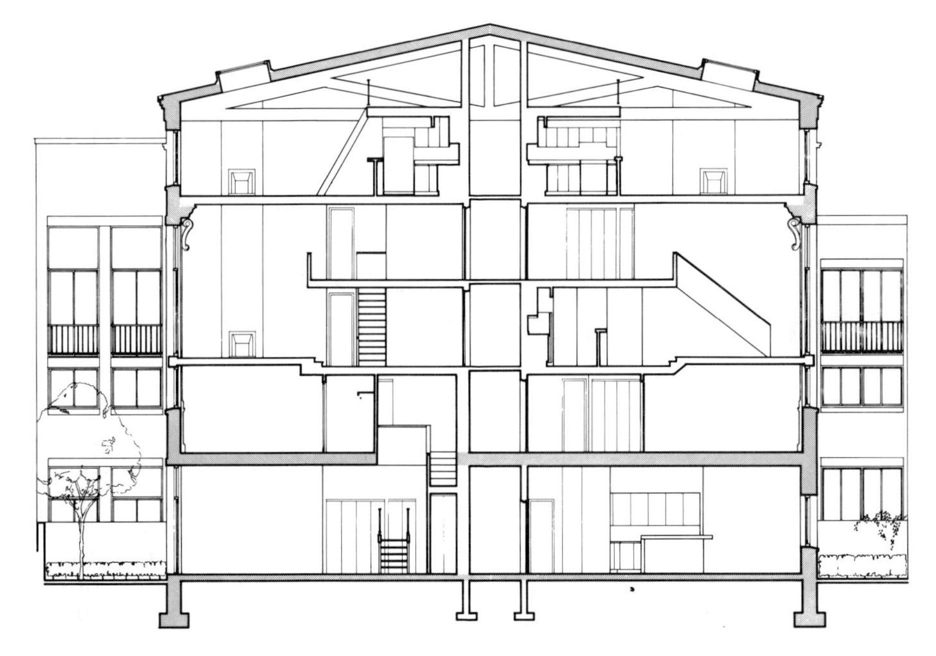

Transverse section. This section indicates how the new floor levels were established within the confines of the existing church structure. Additional usable space was found by lowering the level of the former sanctuary ceiling to the top of the old window casings, and by expanding upward into the roof trusses of the former attic.

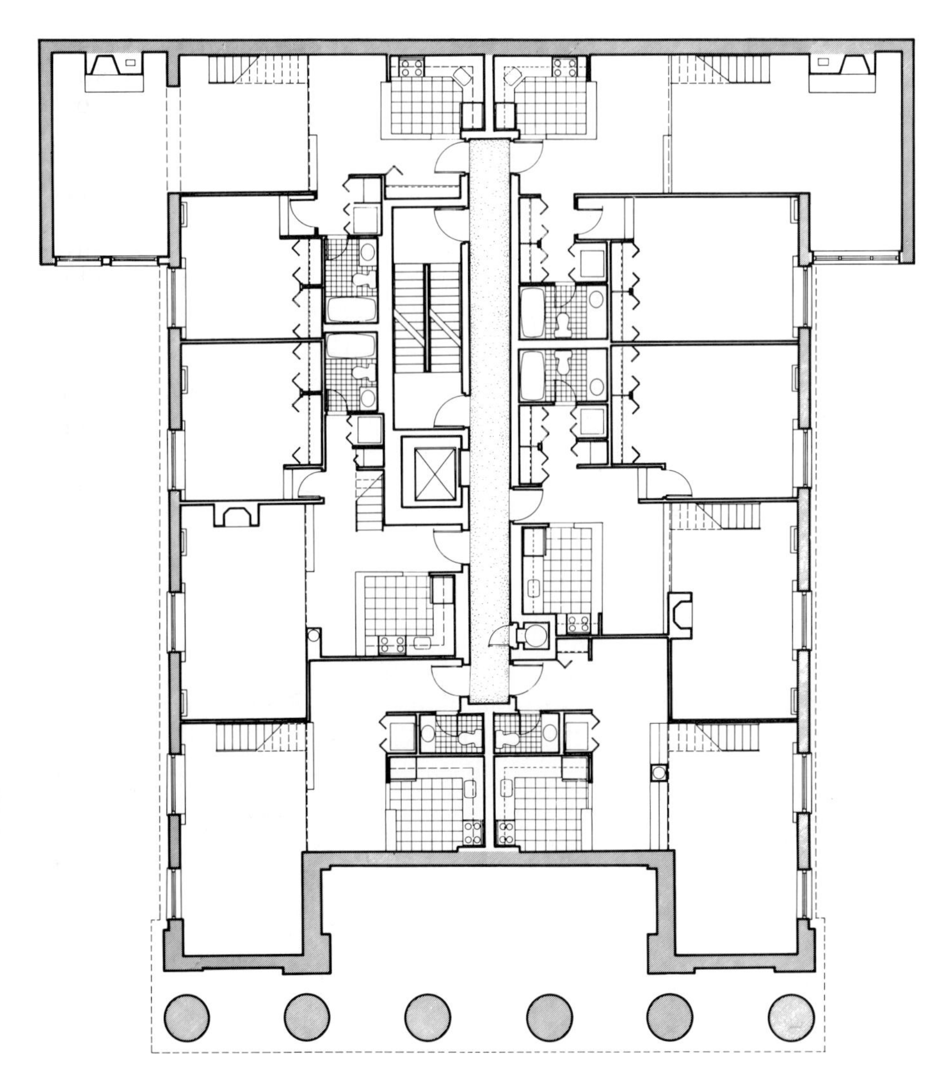

Typical floor plan. This symmetric plan was designed along the axis of a new fire-rated central corridor, incorporating a new elevator shaft and scissor-type exit stairs. The plan and partition layout have been largely determined by the location of existing window openings. All kitchens and bathrooms have been located in the interior to preserve the limited window area for living rooms and bedrooms.

These programs acknowledge that historic rehabilitation costs more than conventional new construction, and are designed to encourage high quality work by allowing investors to recoup a substantial amount of their investment quickly.

Project Approvals

All exterior design work had to receive approval from the New York City Landmarks Preservation Commission. If the developer had elected to take advantage of the tax incentives, additional approvals from the New York State Division for Historic Preservation and the U.S. Department of the Interior would have been required. The federal standard also requires the preservation of significant interior features. Many developers avoid the process because if any aspect of a project is done without permission, delays in construction, fines, withholding a final certificate of occupancy, and a loss of the benefits of the federal tax incentives may result.

Exterior Work

The apartment project was one of the first to be developed under a landmarks review process that included a contract to ensure that the facade will be well maintained in the future.

Much of the previously painted brickwork was restored with a chemical wash combined with high-pressure steam cleaning. The joints were repointed with a mortar and color approved by the commission. Similarly, the commission required that the original bluestone paving in the building's sideyards had to be retained, thereby allowing repair or replacement only where necessary.

The portico facade was repaired where necessary, and the severely deteriorated stucco-faced brick columns were resurfaced. The wood doors, cornice, and pediment were scraped, repaired, and repainted. Stained glass panels on the door transoms were replicated and converted to illuminated light boxes.

Because of the church's monumental presence, the landmarks commission prohibited alterations to the principal street facade. The original church portico was solid masonry except for the main three sets of entrance doors. Restricted by the landmarks commission's requirement that the street facade not be altered, and by his own conviction that reusing the original church entrance doors was symbolically inappropriate, the architect shifted the new apartment entry to a private, paved side court at the basement level. Located behind the blank portico facade are the service cores of the two end apartments.

The absence of windows on the front seemed, at first, to be a formidable obstacle to recycling the church into a residential building. By law, in new or recycled construction, fresh air and natural light must enter windows that are no less than 10 percent of the usable floor area of all habitable spaces such as bedrooms and living rooms. The only acceptable place to provide the required windows in the church was on each side of the building where there is a 12-foot-wide yard or mews. The apartment layouts, some designed as duplexes, were dictated by the desire to utilize the existing tall window openings where possible. To preserve the templelike silhouette from the street, skylights rather than dormers were created flush with the planes of the gabled roof in the new attic spaces where no windows previously existed.

Sideview before restoration *(Far left)*. This close-up view of one of the sideyard alleys prior to restoration shows the original tall window openings that restricted planning the internal layout.

Sideview after restoration *(Left)*. This view illustrates the inconspicuous manner in which the old window openings were modified to suit the new interior layout and the changed floor levels. A new series of windows in the former attic space was created and directly aligned with the existing openings. These slight modifications are restricted to the sideyard alleys and do not detract from the subtle dignity of the Greek Revival columned portico facing the street.

View from street. This photograph illustrates how inconspicuously the former Presbyterian Church has been transformed into Portico Place apartments. The dignified Greek Revival templelike form retains its distinctive architectural presence in the picturesque Greenwich Village streetscape.

Interior Work

Having established the new window and entrance locations, the architect's next step was to fit the new apartment layout within the original volume of the church. The original attic space in the gabled roof was formed by seven 7-feet-deep timber trusses spanning between the exterior masonry walls. The former sanctuary floor, above the basement, consisted of a timber girder-and-joist system, supported by cast-iron columns.

The clear space between the first floor and the low point of the roof is 34 feet, which was divided into four floors. Combining this space with the basement, the developer obtained a five-story building. The first and second floors contain five duplex apartments, four of which have two bedrooms while the fifth has three.

The contractor removed all interior structure, except for the cast-iron columns supporting the first floor, and installed steel floor beams supported on new piers built against the existing masonry. For code purposes, concrete block walls form the public corridors compartmentalizing the building, thus permitting the retention of wood floor framing and roof trusses. A plywood subflooring was installed over metal "C" joists, and oak tongue-and-groove flooring was laid in the apartments.

In the public spaces, precast concrete planks spanning between the concrete block fire walls were covered with carpet. New concrete foundation footings were cast in place at the elevator shaft, and stairway and hallway enclosures. The existing wood roof trusses were reinforced by new wood cord stiffeners.

Fortunately, the building is not oddly shaped, so egress and fire code requirements were easily satisfied. Scissor-type exit stairs along the central spine serve all floors and extend to the roof. The elevator bypasses the first and third floors, which are the bedroom portions of the duplexes. The fire exit stairs are accessible from all floors.

Apartment interior (former sanctuary). Even though it was necessary to carve up the former sanctuary space to create the apartments, retaining an elegant fragment, such as this handsome old bracketed window valence, becomes a decorative focal point in a new living room. Because the room is high ceilinged to permit a balcony on the opposite side, the bold scale of the carved ornament is not overpowering.

Apartment interior (attic space). An interior of striking charm and picturesque character has been created within the former attic space. The exposed trusses add height and interest that compensate for the lack of view and other amenities that a more conventional apartment might provide.

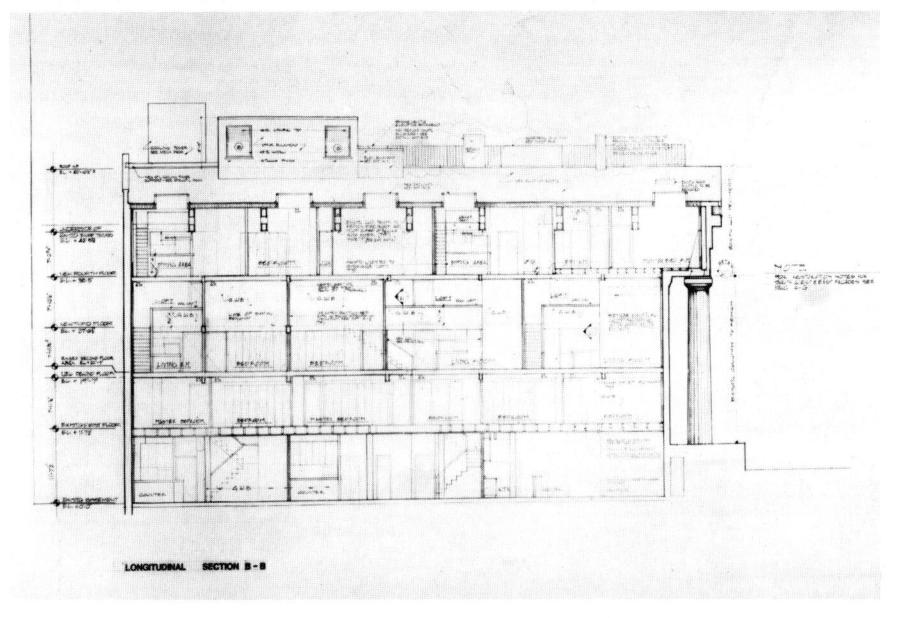

Longitudinal section. This cutaway view demonstrates how the architects have managed to fit a variety of apartments, both duplex and simplex, within the confines of the original church space as mandated by the New York City Landmarks Preservation Commission.

Sprinklers were installed in the building lobby, all public halls, the laundry and compactor rooms on the first level, and in the living and bedroom spaces on the top floor. These apartments are sprinklered so that the wood trusses could be left exposed as a dramatic interior feature. New York City requires that smoke detectors be installed in every apartment.

Two types of units provide central heating. The interior spaces are heated by a hydrotherm modular cast-iron boiler. Finned-tube radiation at the baseboard is provided throughout the building to areas not served by the modular units. These units are reversible, providing cooling in the summer.

Comments

The distinctive, classic, templelike silhouette of this former Greek Revival style church with its fine, columned portico has been handsomely preserved as part of the picturesque residential streetscape of an early nineteenth-century urban historic district.

CHARLES STREET MEETING HOUSE

PROJECT
Charles Street Meeting House
Boston, Massachusetts

OWNER
Charles Street Meeting House Associates

ARCHITECT
John Sharratt Associates Inc.

STRUCTURAL ENGINEER
Brown, Rona Inc.

ELECTRICAL ENGINEER
Zax Associates

MECHANICAL ENGINEER
Environmental Design Engineers

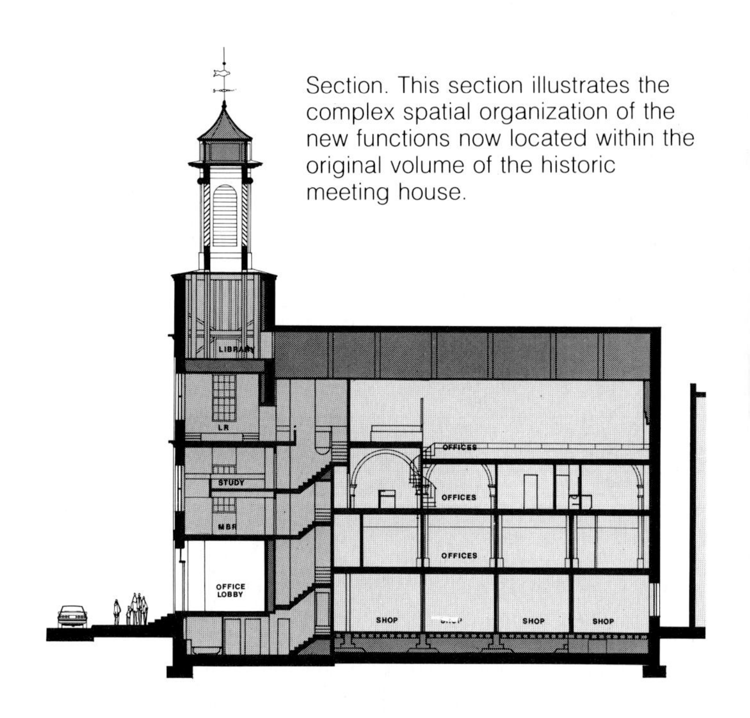

Section. This section illustrates the complex spatial organization of the new functions now located within the original volume of the historic meeting house.

BUILDING SECTION A–A

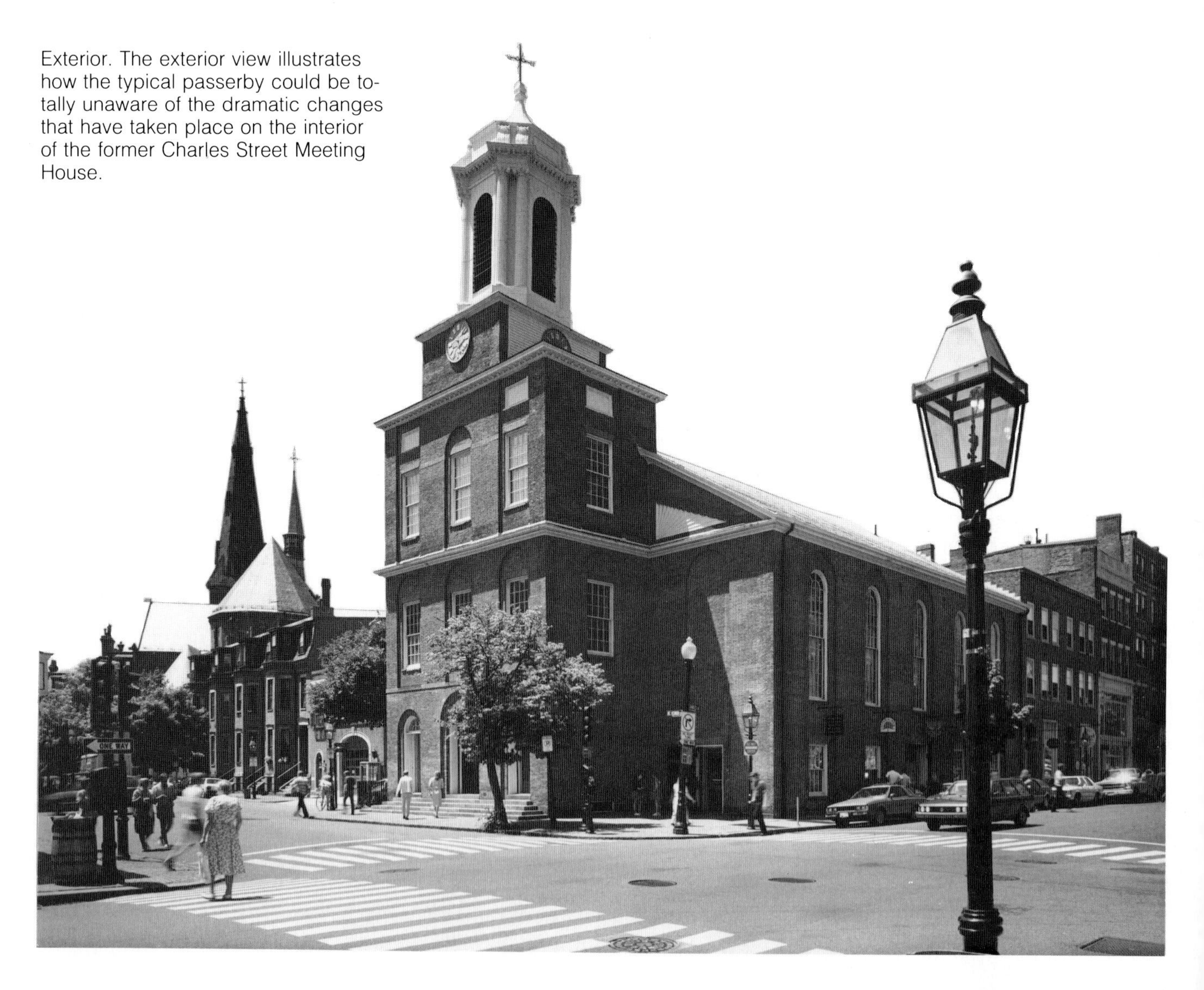

Exterior. The exterior view illustrates how the typical passerby could be totally unaware of the dramatic changes that have taken place on the interior of the former Charles Street Meeting House.

RIVER ST.

GROUND FLOOR PLAN

CHARLES ST.

Ground floor plan. The street level has been divided into a series of small shops, most of which share a common entry vestibule. This arrangement may put the individual merchants at a slight competitive disadvantage, but it was necessary to preserve the original exterior appearance of the old meeting house.

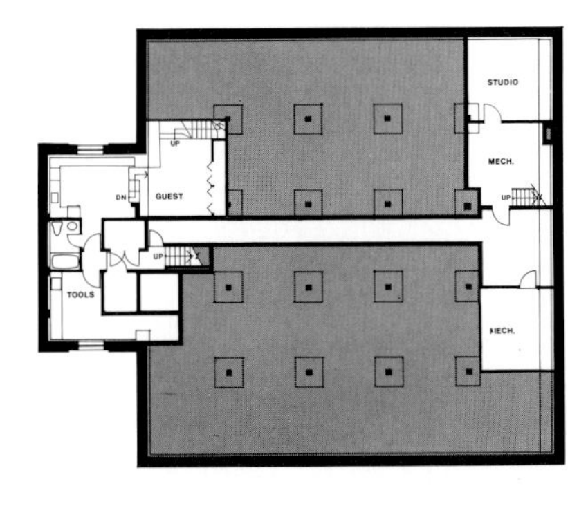

BASEMENT FLOOR PLAN

Basement floor plan. The existing basement, which was not fully excavated, was only partially developed to provide mechanical space and a small guest or caretaker's quarters.

SECOND FLOOR PLAN

Second floor plan. The former second level of the sanctuary now has a flat normal-height ceiling. This level has been divided into small office spaces. New exterior fire stairs have been created at the rear. The bedroom spaces in the former clocktower begin at this level and wrap around a small private elevator and stair circulation system independent of those serving the offices.

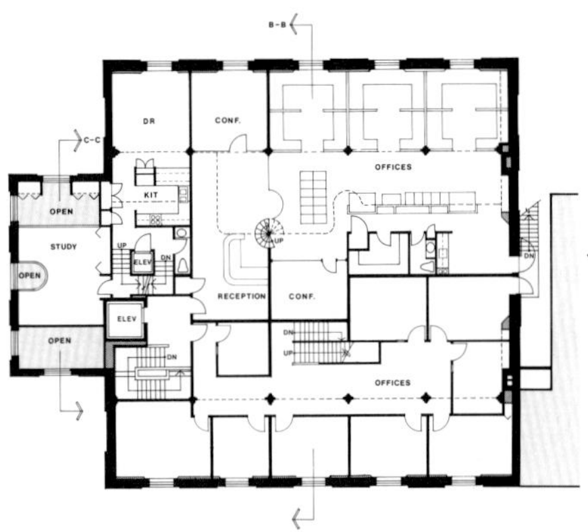

THIRD FLOOR PLAN

Third floor plan. At this level the round-topped arches of the side aisles and the decorative ribbed plaster ceilings of the old sanctuary have dictated the rhythm and layout of the spaces. The partial fourth floor functions as a balcony reached by open stairs, allowing vistas of the lofty volume overhead.

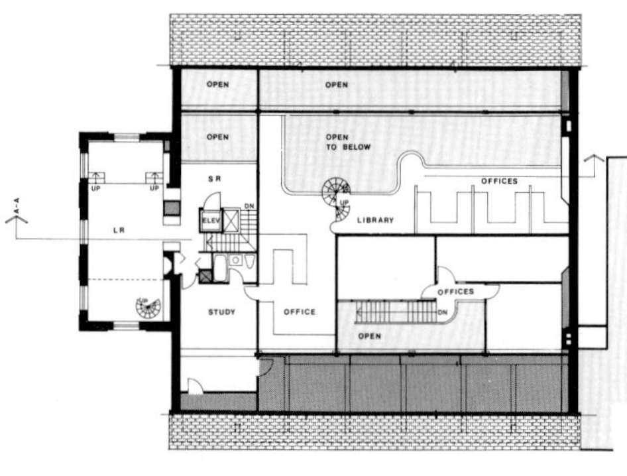

FOURTH FLOOR PLAN

Fourth floor plan. The spaces at the partial fourth floor level take full advantage of the undulating sweep of the old ribbed plaster vaulted ceiling. The living room within the former clocktower and the study connect directly to a personal office within the adjacent architectural offices, library, and drafting rooms.

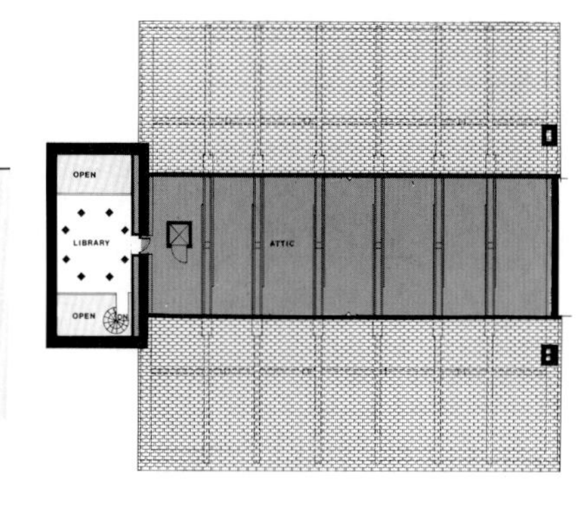

LOFT PLAN

Loft plan. The topmost level of the architect's personal apartment is a small private library containing the framing of the upper-staged campanile of the former clocktower. There is also access at this level to the attic space within the wood roof trusses.

Interior. This interior view (third floor architectural office and drafting spaces) illustrates how much more exciting and enjoyable these working areas are because of their unique setting within the old sanctuary.

HAZELTON HOUSE/ OLIVET BUILDING

PROJECT
Hazelton House/Olivet Building, 1890/1972
Former church, Yorkville, Toronto, Canada

ARCHITECT/DEVELOPER
Sheldon Rosen

Principal entrance. The corner tower at the street intersection has been renovated as the principal entrance to the shopping complex with stair circulation to reach various levels. New graphics, signs, and display windows have been carefully designed to be as unobtrusive as possible.

Exterior. A Richardsonian Romanesque church is prominently located in a residential neighborhood where the surrounding row houses have been transformed into art galleries and boutiques. The former church has been tastefully converted into a multilevel shopping complex that preserves its distinctive architectural detailing inside and out. The volume of main sanctuary space was divided horizontally, retaining the magnificent dome with its semicircular stained glass transom as the top floor.

MANSIONS ON LARGE ESTATES

Changing times and lifestyles have rendered many grand residences obsolete. Soaring fuel costs and maintenance have forced many families and institutional owners to abandon mansions, auction their contents, and sell their estates for development. The successful adaptation of these properties to new uses involves not only an imaginative architectural solution but also extensive negotiations for zoning and building code variances to secure the necessary approvals and adequate financing.

The problems of saving almost identical mansions on large estates (originally designed by the same architect) will be very different depending on what has happened to their surroundings over the years. For this reason Whitefield, located in the oceanfront resort community of Southampton, New York, makes an interesting comparison with Beechwood, overlooking the Hudson River in Scarborough, a prosperous commuter suburb of New York City.

GOLD COAST MANSION RESTORED FOR ELEGANT RESIDENTIAL COMMUNITY

Existing Structure

Renamed "Whitefield" by the developers, the Breese Mansion was originally built for one of Stanford White's personal friends, James Breese, who continued to add rooms, wings, and landscape features until he sold the estate shortly before the 1929 crash, ironically, to the Mr. Merrill, of Merrill, Lynch, Fenner & Smith. At Merrill's death the mansion was left with an endowment to Amherst College as a conference center. Amherst found it impractical to operate and sold it in the 1960s to The Nyack School for Boys, which added a gymnasium and dormitories to the existing estate structures. The school went bankrupt in 1972 and despite numerous development proposals remained vacant until it was acquired by architect Simon Thoresen and his associates in 1979.

WHITEFIELD

PROJECT
Whitefield
Southampton, New York

ARCHITECT
Simon Thoresen & Associates
(now Sculley, Thoresen & Linard)

LANDSCAPE ARCHITECT
A. E. Bye Associates

STRUCTURAL ENGINEER
Paul Gossen

MECHANICAL ENGINEER
Henry M. Fox & Associates

Mansion—Unit 3. Residence 3 has been carved out of the former kitchen service wing of the main house. An elegant pergola provides a private outdoor space for the exclusive use of this unit. The architects have skillfully designed the five individual units with a sense of identity and privacy, without sacrificing the dignity and integrity of McKim, Mead & White's original sprawling turn-of-the-century Colonial Revival mansion.

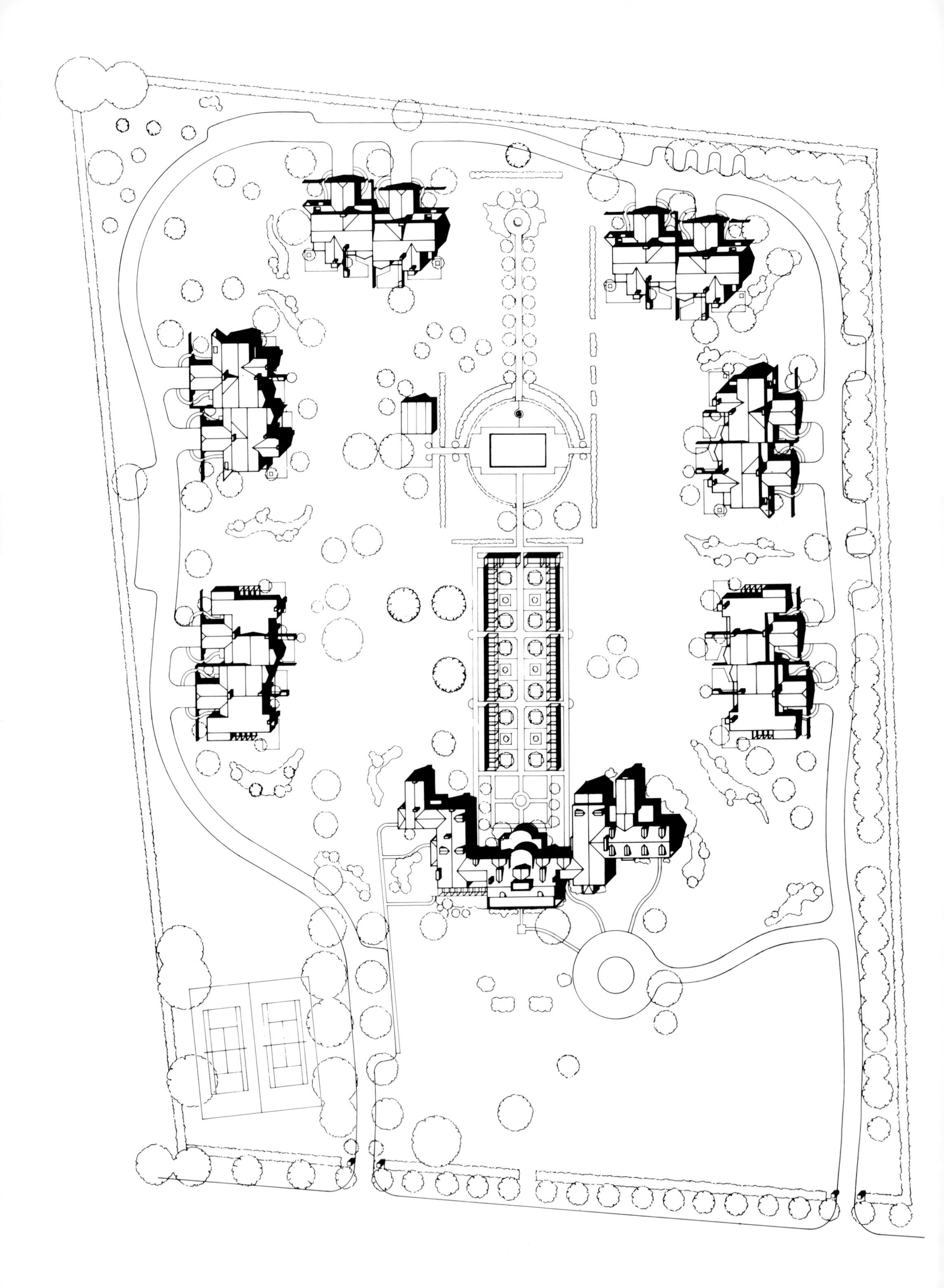

Restoration Program

The architects joined forces with a local realtor and a financier and undertook the project themselves as a condominium cluster development. After a protracted zoning battle and some difficulties securing the necessary financing, the mansion was converted into five residential units and 24 town houses.

Project Zoning and Approvals

In order to maintain the architectural integrity of the exterior of the mansion, as well as to make the venture economically viable, the architects planned to divide the main house into five independent residential units and to build 24 new town house units in clusters within the landscaped 16-acre property. This decision meant that all other existing estate structures, as well as new facilities added by the boarding school, were demolished or removed to other sites. The total of 29 units is the equivalent of one-half acre per single family dwelling, the ratio normally permitted by the zoning ordinance.

The process of planning, code compliance, and construction were no more difficult than similar projects with old buildings, but the maneuvering to achieve the zoning approvals prior to construction proved to be extremely difficult. The village of Southampton is the most fashionable of Long Island's Hamptons Gold Coast resort communities. In this exclusive summer colony, many estates with mansions the size of Whitefield are still staffed and maintained in the traditional manner. Most are discreetly hidden behind 12-foot-high clipped hedges, surrounded by manicured lawns and specimen trees and plantings. Whitefield, which absorbed the original nineteenth-century farmhouse within its sprawling layout, was one of the earliest of the great houses

Site plan *(Opposite)*. The site plan indicates the location of the mansion and the six residential cluster units surrounding the formal gardens. The original entry road on the main axis of the mansion has been eliminated and replaced by a new loop road at the periphery of the site. The cluster units at the rear open toward the landscaped lawns, formal gardens, and the vista of the mansion in the center of which is a new swimming pool. The old tennis courts have been kept and refurbished. All of the old major trees, formal clipped hedgerows, and plantings have also been retained.

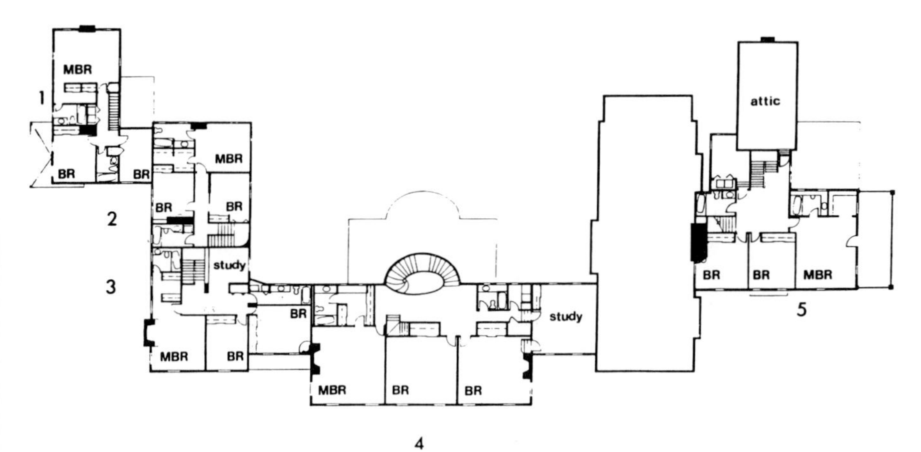

Mansion—second floor plan. On the second floor, in order to accommodate all of the bedrooms, necessary closets, and bathrooms, very little of the original layout has been preserved. The bedrooms are contemporary and have generous sliding door wardrobes. The original fireplaces have been retained.

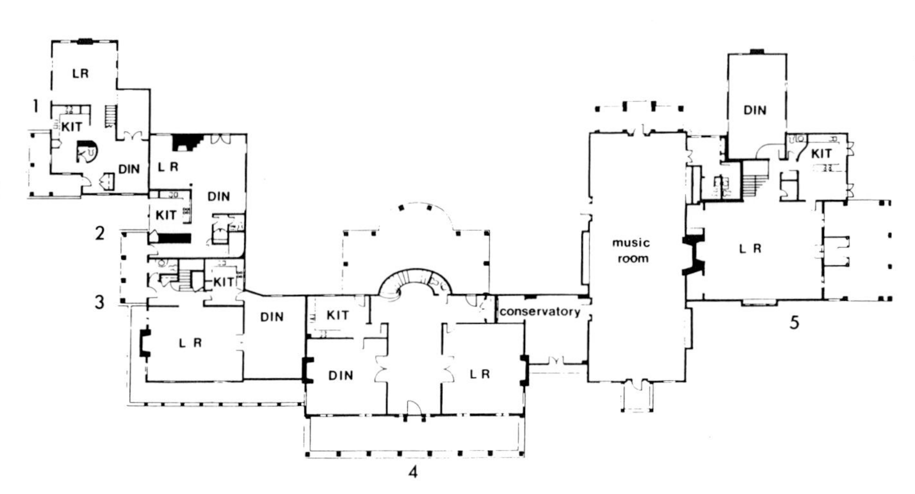

Mansion—first floor plan. The plan indicates the division of the main house into five independent duplex residences of varying sizes, all entered from an individual porch or portico. Units 3, 4, and 5 have incorporated some of the principal rooms of the mansion, preserving much of their original architectural character and decorative features. The enormous original Music Room and the adjacent conservatory have been retained for the common use of all the condominium owners at Whitefield.

built in Southampton. Unfortunately, its inland location is isolated from the more desirable beachfront property favored by later arrivals. Earlier subdivision of the original Breese property of about 100 acres resulted in modest houses on postage-stamp lots bordering the estate. Changes in zoning from residential to institutional classification for Amherst's Merrill Center for Economic Studies, and subsequently for The Nyack School for Boys, complicated the situation. The local government had never previously considered the prospect of dividing a large single-family residence into multifamily or apartment use and was terrified of setting such a precedent.

Further, the architects and developer were "outsiders" from New York City, and they proposed building condominiums, a type of development the town fathers associated with more flamboyant resorts in the Sun Belt. Consequently, the somewhat suspicious village board dragged its feet. Meanwhile, the financial institutions thought that the projected costs of the renovation and construction would make the units too costly for the second-home market. They also wanted assurances that the village would grant the necessary zoning variances for the project.

The village, taking a similarly cautious position, wanted to see the financial commitment before granting the zoning variance for multifamily cluster development. Additionally, the village was still concerned about setting a precedent that would enable other developers to subdivide large houses into small units. The easiest course would be to

Mansion—elevations. The elevations illustrate the sprawling character of the Whitefield mansion, which sprouted many wings, extensions, and accretions representing a collaboration between its original owner, James Breese, and his architect friend, Stanford White.

maintain the mansion's institutional occupancy zoning laws dating from its days as a study center and boarding school.

To help move the proceedings, the architect suggested that the village board issue a special zoning permit for multifamily use, thereby retaining its control of future projects on a case-by-case basis. Then the architect/developer cooperated to obtain the property's listing on the National Register of Historic Places, which assured many local critics and village officials that the architectural integrity of the mansion would be respected.

Exterior Work

The designation as a multifamily structure automatically put the project into a fire district. Normally, this means that the building would have had to have noncombustible walls, compromising its white clapboard and shingle, Colonial Revival exterior. Because of its National Register status, the architect was able to obtain a dispensation from New York State fire code enforcement officials to retain the original exterior wall finishes, provided that a sprinkler system and other fire preventive measures were installed. The new construction, designed to complement the old mansion in scale and materials, also had to be fireproofed and sprinklered.

Exterior walls were originally clad in oversize cedar shakes exposed 16 inches to the weather. Since replacements would have had to have been custom fabricated, the shingles were stripped of old paint and repainted white. The deteriorated roofing was completely replaced with new fire-retardant cedar shingles.

Cluster house—entrance facade. A typical cluster house containing four residential units seen from the new peripheral loop road. The architects have designed these units in a contemporary idiom echoing the traditional gabled roof profile of the Colonial Revival mansion.

Cluster house—garden facade. The garden facades of the cluster house are full of windows opening toward the main axis of the formal landscape and gardens. Each of the four residential units has been provided with a private terrace and outdoor space. Visually differentiating the private areas from the common space, without the need for fences or barriers, is essential to preserving the luxuriant parklike quality of the traditional landscape.

Interior Work

Fire walls were built as a vertical separation between the five residential units, and where one unit horizontally overlapped another, the floor between them was filled with lightweight vermiculite insulation. As a further precaution against the spread of fire, the basements were not directly connected to the living units and were accessible only from an exterior entrance; also, the attic spaces could not be occupied. Sprinkler systems were installed in the attics, basements, halls, and stairs to provide safe exit for the occupants.

Structural work included the relocation of some load-bearing partitions and installation of new stairs. One of the most ambitious feats was eliminating a sagging portion of the floor in the main house. The floor was jacked up from beneath to a level alignment while plywood underlayment was glued to the joists to make them more rigid. Then new fire-rated walls separating the individual residential units were installed. When the jacks were removed the trussed walls carried the floor-load without deflecting.

Each residential unit has an independent hot water heating system connected to its own boiler in the basement. High-velocity air conditioning with a flexible plastic-duct system was chosen because it could be maneuvered to fit difficult configurations. (The attics were convenient for running extensive duct systems.)

Music Room interior. The great Music Room has been restored as a common space for the use of the Whitefield residents. Stanford White saw nothing inconsistent in encasing this eclectic fantasy in a white clapboard colonial revival wing.

Special Interiors

Although the mansion has an impressive exterior and some superior eclectic interior spaces, the majority of the rooms were quite plain. The decor of the Etruscan Room, designed for Breese's collection of classical objects, has been preserved as a dining room in one of the units. The lattice-lined conservatory, once a porte-cochere, has been preserved as the entrance to the mansion's Music Room. When the upstairs floor plans were redesigned, no special efforts were made to preserve or duplicate any moldings or decorative embellishments. Only the ground floor had wall paneling relocated, and there was scant molding or trim of distinctive character on that floor.

The most extraordinary room in the mansion is the Music Room, which is a Stanford White fantasy. One enters from the conservatory, through a Federal style doorway with a leaded fanlight and sidelights, into a vast 30 × 70 × 20 foot space finished with medieval oak linen-fold paneling, a painted Italian Renaissance ceiling, a French limestone Renaissance fireplace chimney, and bay window with stained glass depicting the coats of arms of the signers of the Magna Carta. A built-in Aeolian organ console and a gallery for the organ pipes remain intact. This remarkable interior and the conservatory have been restored and furnished for the common use of the 29 owners of Whitefield.

Gardens

An important element of the project was the restoration of the formal gardens that were a prominent feature of the original estate. After several years of abandonment and neglect the beds, parterres, and hedges were overgrown and in total disrepair. The vines were untangled from the pergola and not only did the wood trellis require replacement but the cast-stone columns as well. Because the property is listed on the National Register of Historic Places this restoration had to be carefully executed. Russell Page was the consultant for what turned out to be a monumental effort. The choice of plantings reflects the desire for less elaborate maintenance than the original gardens. The splendid results provide a visual focus for all of the residents of Whitefield.

Comments

The success of this project, which has been widely publicized, led to the Beechwood project in Scarborough. Its example has yet to be followed in the nearby Gold Coast resorts of Long Island, which remain resistant to any modifications that might set a precedent for multiple units or apartments.

Formal gardens—pergola. The old pergola defines the formal parterre gardens on the rear axis of the mansion. Once the restoration was begun it was discovered that the masonry columns supporting the wood pergola were so deteriorated that entire structure had to be rebuilt.

View from mansion. This view from the rooftop of the mansion shows the white painted clapboard wing containing the Music Room. In the distance the new cluster units surrounding the formal gardens echo the architectural form and detailing of main house.

BEECHWOOD

PROJECT
Beechwood
Scarborough-on-Hudson,
New York

ARCHITECT
Simon Thoresen & Associates
(now Sculley, Thoresen & Linard)

LANDSCAPE ARCHITECT
A. E. Bye Associates

STRUCTURAL ENGINEER
Paul Gossen

MECHANICAL ENGINEER
Henry M. Fox & Associates

NEW RESIDENTIAL COMMUNITY ON COUNTRY GENTLEMAN'S ESTATE

Existing Structure

A suburban estate consisting of a rambling, wood turn-of-the-century mansion and its mature landscaped grounds remained empty and unsold. Beechwood typified the dilemma facing many older suburbs whose once-proud mansions are threatened by neglect and abandonment while, at the same time, a shortage of open land and housing is creating tremendous pressure for speculative development. Beechwood is a sprawling clapboard Colonial Revival mansion that incorporates a 1790 house within its nineteenth-century additions. It was designed by the well-known architects R. H. Robertson and Wells Bosworth. Mr. Bosworth also designed the swimming pool, formal gardens, and various landscape structures for a prosperous New York City banker. The house had a long and flamboyant history before falling into disrepair.

Situated on a hilltop overlooking the Hudson River, it is a surviving link in the continuous chain of estates that once stretched along the river's eastern bank all the way to the state capital in Albany. In the nineteenth century the railroad was located along the river's edge. The Scarborough commuter railroad station, 45 minutes to Grand Central Station, lies at the foot of the hill beneath Beechwood's river vista.

Site strategy. The small site plans illustrate the analysis made by the architects in determining which original structures and landscape features would be preserved and integrated with new construction in adapting Beechwood from a mansion to a multifamily condominium. Old barns, stables, and greenhouses were demolished to reduce maintenance. Major elements of the original mature landscape, road network, formal flower gardens, reflecting pools, pergolas, and the swimming pool were retained and restored.

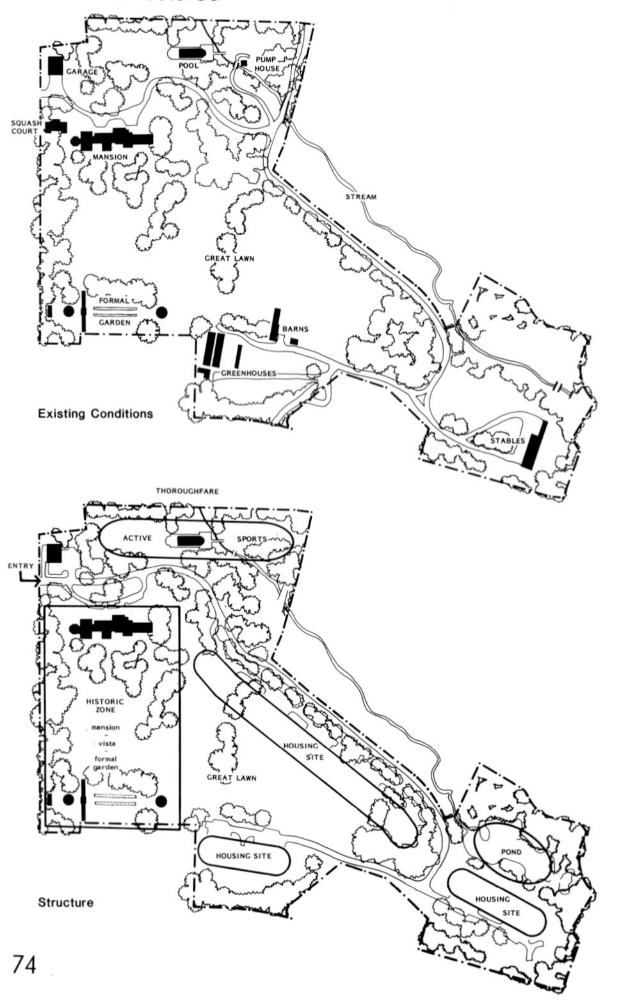

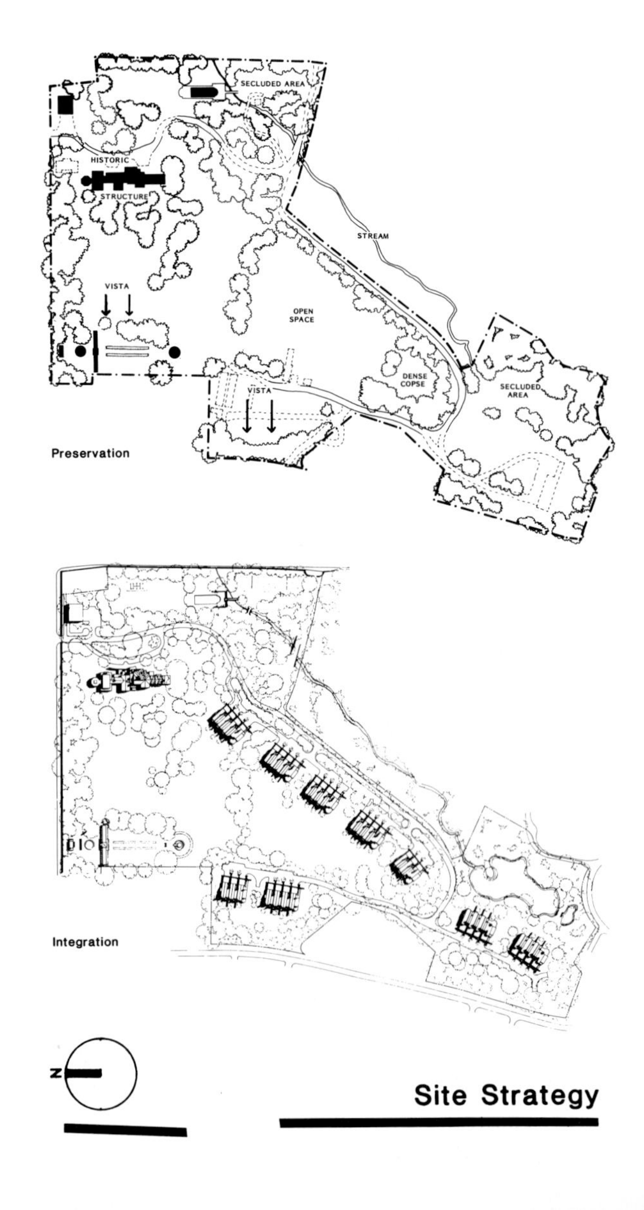

Site Strategy

Mansion elevations. As the major elevations indicate, the old mansion evolved over time and was not symmetrical. This facilitated the architects' efforts to provide distinctive individual entrances to the three new residential units and the common space into which it was divided.

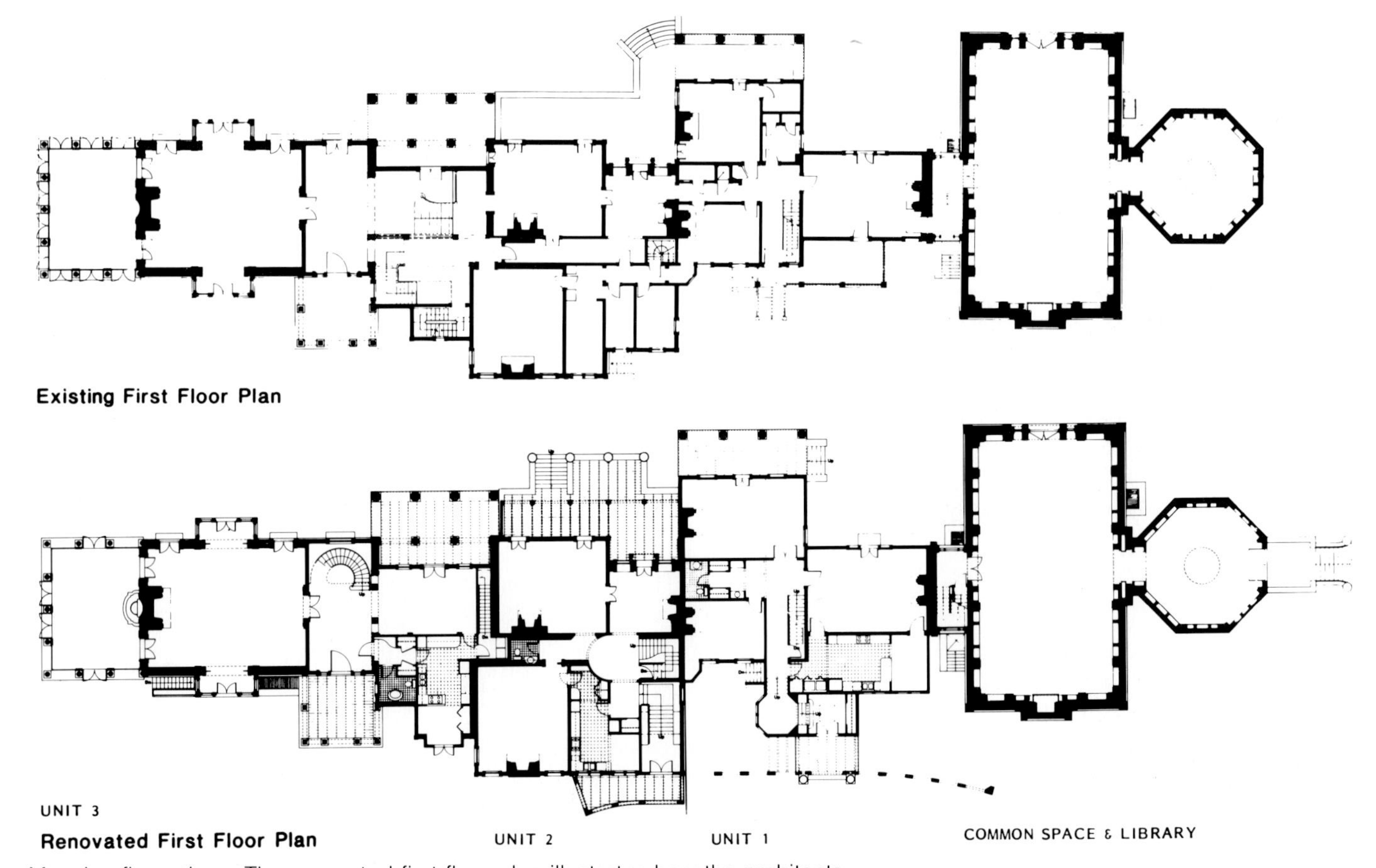

Mansion floor plans. The renovated first floor plan illustrates how the architects have divided the existing spaces into three residences. Major original rooms and fireplaces were integrated into the plans. The large rectangular original library and its octagonal annex were retained for the common use of all the residents of Beechwood. The major staircase and a warren of minor accessory spaces, small service corridors, and closets were demolished to provide new kitchens, guest lavatories, stairs, and circulation spaces for the three units.

New cluster units—road elevation. The architects have preserved the old estate road network and lined it with mature trees. The garages and fenced small entrance courts assure privacy for the new clustered residences.

Forming the inland boundary of the property is a historic stonewall-bordered road once lined with the gate houses of the great estates but now a highly traveled artery. The surrounding properties on the descent to the railroad station have now been rezoned for attached town houses and apartments, setting a pattern for the breakup of Beechwood's acreage, which would have normally yielded a single family house for each three-quarters acre. The formally landscaped grounds of the estate included a teahouse, a pergola, giant classic columns, and a swimming pool.

Restoration Program

The house had been neglected for several years and the neighbors considered the property their own private park. When plans were announced for developing Beechwood, the neighbors vigorously opposed the rezoning necessary to develop it. The architect was able to convince them that cluster houses would preserve the parklike quality and promised not to destroy any major trees or build any houses directly in front of the mansion. These promises were kept and the mansion retained its breathing space.

The architect joined forces with a financier to develop the 34-acre site and market the housing units. The mansion itself was converted into three substantial residences and the additional 34 houses were clustered to preserve as much open space as possible.

New cluster units—garden elevation. The design of the new cluster units orients them toward the great rolling lawns and greenswards of the old estate landscape. Small individual outdoor sitting areas are defined by wood walls, clearly separating private from communal spaces. To preserve the traditional parklike sweep of the estate's grounds, the fences and hedge walls have been held to a minimum.

Project Approvals

Planning for the project proceeded smoothly because Scarborough is in the jurisdiction of the village of Briarcliff Manor, which has an enlightened planning board. The board said, in effect, "Show us the preliminary proposal and we will indicate whether or not development can continue." The proposal met with the board's unofficial approval, and the rest of the planning went on without the antagonism frequently encountered in similar suburban residential development proposals.

One advantage that facilitated the Beechwood conversion was its location in a free-floating zoning district, which enabled the planning board to assign it any occupancy they deemed suitable. This meant that the building is not considered to be in a fire district, and no dispensation was required for leaving the original walls in their original material.

Subsequent to completing the project a change in the New York State Building Code allowed the third floor of an existing house to be used as habitable space. Thus, the new owners were able to expand into the attic spaces, an unanticipated bonus that increased the market value of their units.

Exterior Work

Restoration of the exterior of the mansion is almost a repetition of Whitefield in Southampton. The wall cladding, which was a melange of wood clapboard, shingles, and siding, was stripped and repainted white, and the roof shingles replaced. The design of the clustered-house grouping is similar in character to the one at the Whitefield estate which was developed by the same architect/developer/financier team.

Landscape Preservation

Perhaps even more remarkable than the house itself is the 34-acre site and landscape, which the architect/developer was instrumental in getting included in a historic district. The restoration of the landscaping is important to recapturing the original ambiance of the estate. To assure that the property continues to be well maintained, the 37 residential units were sold in fee-simple ownership, not as a condominium or cooperative, and the buyers were obligated to participate in a homeowners' association that maintains the property.

Mansion portico. The original entrance portico has been retained as the entrance to one of the three new residential units. Distinctive new contemporary interpretations of the classical wood-columned entrance portico have been designed for the other two units.

Interior Work

Because the mansion was the focus of the property, its exterior was not significantly modified, except to make new front entrances where there had been service entrances. The interior was carved into three large, independent residences, ranging from 3,000 to 5,000 square feet, each with its own entry. Interior fire walls separate the units, and individual heating and air conditioning systems were installed.

Beechwood, however, retains many of its original rooms. The grand formal oak-paneled living room was restored along with the dining room, the most significant ground floor rooms. The original main stair was too large for the divided house and had to be removed. A new helical stair, reminiscent of that at Whitefield, replaced it. Wherever possible, old moldings, doors, and hardware were cleaned, repaired, and reused. Marble bathroom fixtures were furnished with new plumbing fittings. Duplicating wood and plaster molding profiles was required, as well as decorative plaster ceiling medallions that were reproduced in fiberglass. The showpiece of the mansion, the library, has been restored and maintained as an amenity available to all the residents of the estate.

Comments

Many large suburban estates have been compromised by the transformation of former country roads into heavily traveled routes; the loss of privacy has made estates less desirable as residences for wealthy individuals. By permitting the conversion of Beechwood to apartments, and the addition of clustered town houses within the landscaped grounds, an important architectural and visual anchor in the suburban roadscape has been preserved.

Shifts in the character and status of once-prestigious communities, as well as changes in lifestyle, have contributed to the abandonment of some of our finest residential properties. Often the surrounding acreage is sold off and inappropriately subdivided so that great houses are left stranded on stamp-sized lots. These fine old houses are further compromised by new neighboring corporate headquarters and office parks that generate traffic jams. Often it is beyond the means or the desire of their owners to offer workable solutions, yet more and more communities have insisted that something must be done to preserve these properties.

HENDERSON HOUSE

PROJECT
Henderson House, Private Corporate Conference Center
Selma, Alabama

OWNER
Circle "S" Industries

ARCHITECT
Nicholas H. Holmes, Jr. (now Holmes & Holmes) and James Seay, Sr. (now Seay, Seay & Litchfield)

LANDSCAPE ARCHITECT
Sanford Chandler

INTERIOR DESIGNER
Helen Sapp Interiors

PROJECT DIRECTOR
Melissa L. Spann

CONSTRUCTION MANAGER
T. Nichol Lux

SOUTHERN MANSION RESTORED TO FORMER GRANDEUR

Existing Structure

In order to restore a landmark pre-Civil War mansion (1853), Henderson House, and adapt it to a new commercial use, it was necessary to stabilize its immediate surroundings, a deteriorating turn-of-the-century residential neighborhood of modest bungalows.

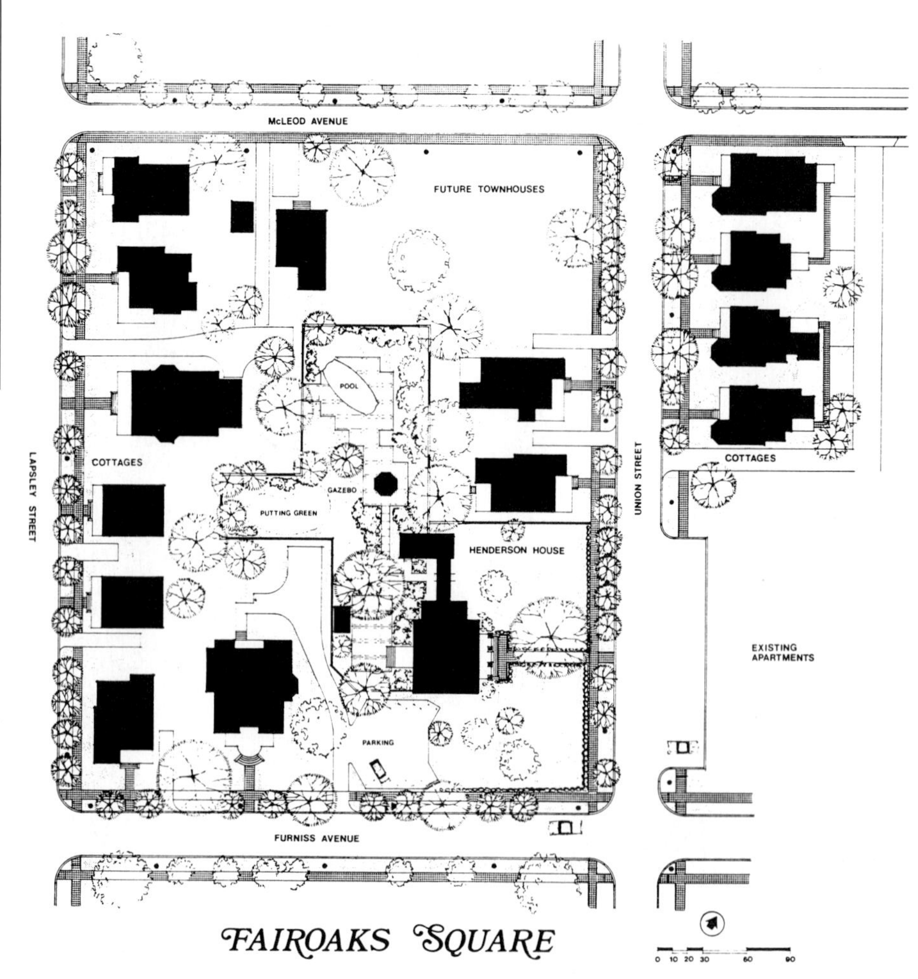

Site plan. The site plan shows the relationship of Henderson House, which has been converted into a corporate conference center and guest house, to the small residential cottages that surround it. The skillful landscaping and placement of the mansion's garden walls at the core of the residential square have enhanced the attractiveness of the setting while providing visual separation and privacy for the neighbors.

Portico restoration *(Above)*. An extensive program of repairs was undertaken to restore Henderson House to functional use. Scrupulous attention was paid to historic accuracy in all aspects of the restoration.

Main portico *(Left)*. This view of the restored portico shows how unobtrusively the mansion-cum-corporate guest house fits into its residential surroundings.

Rear porch *(Above)*. An old photograph gives some indication of the extensive repairs that were necessary to restore the mansion for use as a corporate guest house.

Rear garden *(Left)*. The rear portico has been linked to the former outbuildings by a network of traditional brick-paved paths as part of a new garden incorporating a gazebo, a swimming pool, and a putting green.

Restoration Program

Two Selma officials, Mayor Joe Smitherman and Mrs. Elizabeth Driggers of the city's Office of Planning and Development, had the foresight to rescue the main house and patiently show it to prospective users for several years until Circle "S" was interested. Circle "S," a local manufacturer of architectural products, bought the mansion and 11 cottages on the same block. Concurrently, the company restored the mansion for its own use as a conference center and guest house, and renovated the cottages and resold them for upper middle-class family occupancy. In effect, this became a privately sponsored small-scale urban renewal project not unlike the Duke Foundation's rescue efforts in Newport, Rhode Island's eighteenth-century neighborhoods.

Circle "S" Industries retained architects Nicholas H. Holmes, Jr., a historic preservation specialist, and James Seay, Sr., to restore Henderson House as authentically as possible while adapting it to contemporary corporate requirements. The design of the conversion for use as a corporate conference and guest house facility was fitted into the existing volume of the mansion and its original outbuildings.

Project Finances

The owner acquired the building from the city for $55,000 and spent $600,000 to restore it. The city of Selma contributed about $100,000 from CDBG (Community Development Block Grant) and historic preservation grants and an additional $15,000 in labor and services. The city also applied a portion of a $270,000 Urban Development Action Grant to provide paved brick sidewalks, lighting, street improvements, and the relocation of utilities underground on the public property surrounding the block of restored houses.

Project Approvals

In addition to having to obtain the local building approvals, since the property is listed on the National Register of Historic Places, it was also subject to aesthetic review by the U.S. Department of the Interior for all the modifications and additions to the mansion, the outbuildings, and the site itself.

Exterior Work

Most of the original wood clapboard siding has been retained, but where patching was needed, the contractor supplied a cache of 120-year-old planks for the repairs. Similarly, the original finished flooring was maintained intact except in one room found to have termite infestation. The flooring in this room was replaced with old pine of the same age as the building. Today, as preservationists seek replacement parts from materials depots, the demolition contractors themselves have come to recognize that more careful salvage methods can yield a marketable commodity.

The original terne metal roofing was still in good condition and serviceable. It was carefully cleaned without abrasives, repaired, and repainted.

Special efforts were made to prevent a reoccurrence of the deterioration of the four wood Doric columns supporting the traditional Southern style front portico. During the restoration, the columns were raised while new concrete piers were cast under them. Concealed metal plates were inserted to maintain a quarter-inch clearance so that adequate air circulation would prevent excessive condensation that had previously rotted the bases of the wood columns.

During restoration, crews stripped all exterior and interior paint down to the bare wood. Chemicals removed the lead-, oil-, and water-based layers, but it could not remove the building's original milk-based white paint. This stubborn coating required a strong solution of ammonia for its removal.

Interior Work

All plaster-covered interior portions were scraped and scarified to remove loose material, coated with a bonding agent, and resurfaced with new plaster. Deteriorated decorative plaster, moldings, and coves were recast in plaster and missing wood detailing was reproduced from old wood.

Age had caused the cantilevered wood main and attic stairs to sag. A new welded-steel channel under each stair carriage braced them back to their original positions, and adjustable bolts were provided to compensate for future sagging.

The original kitchen area, as is typical of Southern architecture, was a separate structure linked to the main house by a covered walkway. This area was re-equipped as a commercial kitchen. To provide additional interior space within the original exterior shell, the massive cooking fireplace and chimney were removed. The walkway to the main house has been transformed into an enclosed passageway with an aluminum-framed glazing manufactured by Henderson House's new owner.

Two bathrooms on the ground floor were fitted into an existing single-story service wing on the same side of the house as the kitchen building. To accommodate the program requirements of bedrooms and conference rooms on the second floor, two new bathrooms were added above the service wing. This stacking arrangement provided some economies in the plumbing requirements.

Parlors. Although old and neglected, the original interiors were basically intact and did not require extensive restoration.

Restored parlors *(Below)*. Since the major architectural features and detailing remained substantially intact, the interior restoration consisted of repairing, refinishing, and concealing modern mechanical systems.

Stair before restoration. Skilled carpentry has permitted this elegant stairway to be retained as the principal feature of the central hall.

Restored central hall *(Right)*. The central hall and stair after restoration retains the dignified atmosphere of a house museum while successfully functioning as a corporate guest house.

A new heating, ventilating, and air conditioning (HVAC) system was zoned into two units. The principal mechanical unit is located in the garden and concealed by shrubbery. This system feeds the ground floor through a network of underfloor ducts. An independent unit downfeeds the upper story and is located in the attic. This type of dual-zoned system is more efficient and economical to operate, since it permits unoccupied portions to be shut down. There are further economies, since the ducts can be easily hidden in secondary spaces such as attics, providing a great deal of flexibility in the location of grilles as well as minimizing costly rerouting to avoid structural conflicts.

Comments

This successful combination of public and private initiative has resulted in much more than the rescue of a historic house, since it has become the focus of a "new" residential neighborhood. With corporate usage low key and unobtrusive, Henderson House provides a well-maintained landscape for all to enjoy.

COMMERCIAL DOWNTOWNS

All over America, center cities that survived the urban renewal bulldozer and the flight of commerce to the suburban shopping mall are now struggling to re-establish their identity. Many communities have initiated planning studies to salvage their remaining historic architectural fragments and reintegrate them in a new "downtown" commercial core. Typically, the remaining older buildings are too small to warrant the individual investment required to rehabilitate them.

Linking a cluster of smaller adjacent structures preserves the historic architectural character of the street facades yet offers increased flexibility of layout in the combined interior spaces. The cost of upgrading to incorporate new fire stairs, automatic elevators, and modern HVAC systems is more economical when spread over several buildings. Larger floor areas also permit more imaginative layouts, including schemes that are, in effect, urban "mini-malls." These can compete more effectively with their large suburban rivals than can the traditional small individual downtown merchant. Shared costs of maintenance, security, and advertising are often additional benefits of this approach.

SMALL COMMERCIAL STRUCTURES COMBINED

THE BANK CENTER

PROJECT
The Bank Center
Pittsburgh, Pennsylvania

ARCHITECT, Preliminary Design
IKM Partnership

ARCHITECT
Lorenzi, Dodds & Gunnill

STRUCTURAL ENGINEER
R. M. Phillips

MECHANICAL ENGINEER
Caplan Engineering

Existing Structures
A series of small underutilized turn-of-the-century commercial structures are located within the densely packed narrow streets of Pittsburgh's downtown financial district. They have been rehabilitated as part of a larger planning effort to reinvigorate the faded historic commercial core and attract new businesses and investment.

Restoration Program
This project involved linking the lower stories of five adjacent commercial structures to an "arcade" or indoor mall to house small boutiques, cinemas, and restaurants as amenities for downtown office workers and shoppers. Plans to convert the buildings into retail spaces started in 1973, when restaurants with bare brick walls and abundant plants were still something of a novelty. However, the initial design for transforming the interiors of the five buildings into the one connected space underwent a series of changes to accommodate the budget imposed by succeeding consultants to the developers.

Often city planners are so captivated by the creation of massive new towers, which are dropped into the existing matrix of older smaller-scale commercial structures, that not enough attention is paid to the resultant loss of convenient and affordable street level amenities for office workers, shoppers, and pedestrians. The Bank Center demonstrates the intriguing possibilities of preserving architectural variety and traditional character of older, small-scale structures by linking them to provide an attractive, larger, and more flexible interior than the more competitive "new" downtown.

Exterior Work
Most of the refurbishing is on the inside of the buildings. The load-bearing masonry exterior walls were only cleaned or repointed where necessary, and since there was only minor deterioration, the original facades remained more or less untouched. There was no applicable historic preservation ordinance, but recommendations contained in street elevation guidelines, which were developed as part of a downtown

Exterior. As the focus of downtown renewal programs shifts to new shiny corporate towers many dignified "old timers" are being abandoned. Behind a group of somber Beaux-Arts facades, some neighboring small commercial structures have been linked to provide an urban "mini-mall" or arcade.

The Bank *(Right)*. A once-proud bank has lost its status and is now outflanked by noisy commercial neighbors. Some of its fading luster has been restored by creating The Bank, a complex of small shops and a movie theater in its cavernous, skylit former banking hall.

Pittsburgh planning study, were followed. On the exteriors, canvas canopies, a logo, and coordinated graphics identify the five structures comprising The Bank Center.

Interior Work

Unlike many restoration projects, these five buildings were in sound physical condition. The principal structure, the bank, had only recently been vacated when its original occupant moved to a new building. All of the original decorative features of the marble-lined, tee-shaped, two-storied main banking hall were left intact, including a magnificent set of stained glass skylight panels. The protective skylight enclosures above the decorative stained glass ceiling panels required some restoration to eliminate leaks. The original mosaic tile flooring of the street-level banking floor had worn thin in spots from years of foot traffic. An analysis of maintenance costs indicated that carpeting would be the most economical new floor surface. Therefore, the mosaic tile was left intact and covered with wall-to-wall carpeting.

An elevator was installed to serve the newly created second story and basement commercial spaces. Costs for this and other mechanical improvements became more economical when spread over the five buildings. Sprinklers were added, and the central heating and air conditioning systems were revamped. The new ducts, pipes, and conduits were run exposed in keeping with the designer's aesthetic that quite deliberately contrasted the new installations against the existing formal interior finishes and detailing. It was the developer's intention to allow the individual mall occupants to design their own retail settings within the monumental interior environment provided by the former banking hall.

Comments

The Bank Center project has preserved a handsome but neglected block of Pittsburgh's traditional downtown financial district without public financing. In the process, it has stabilized five old buildings and provided ongoing maintenance so that they may continue to adapt to the changing demands of commerce.

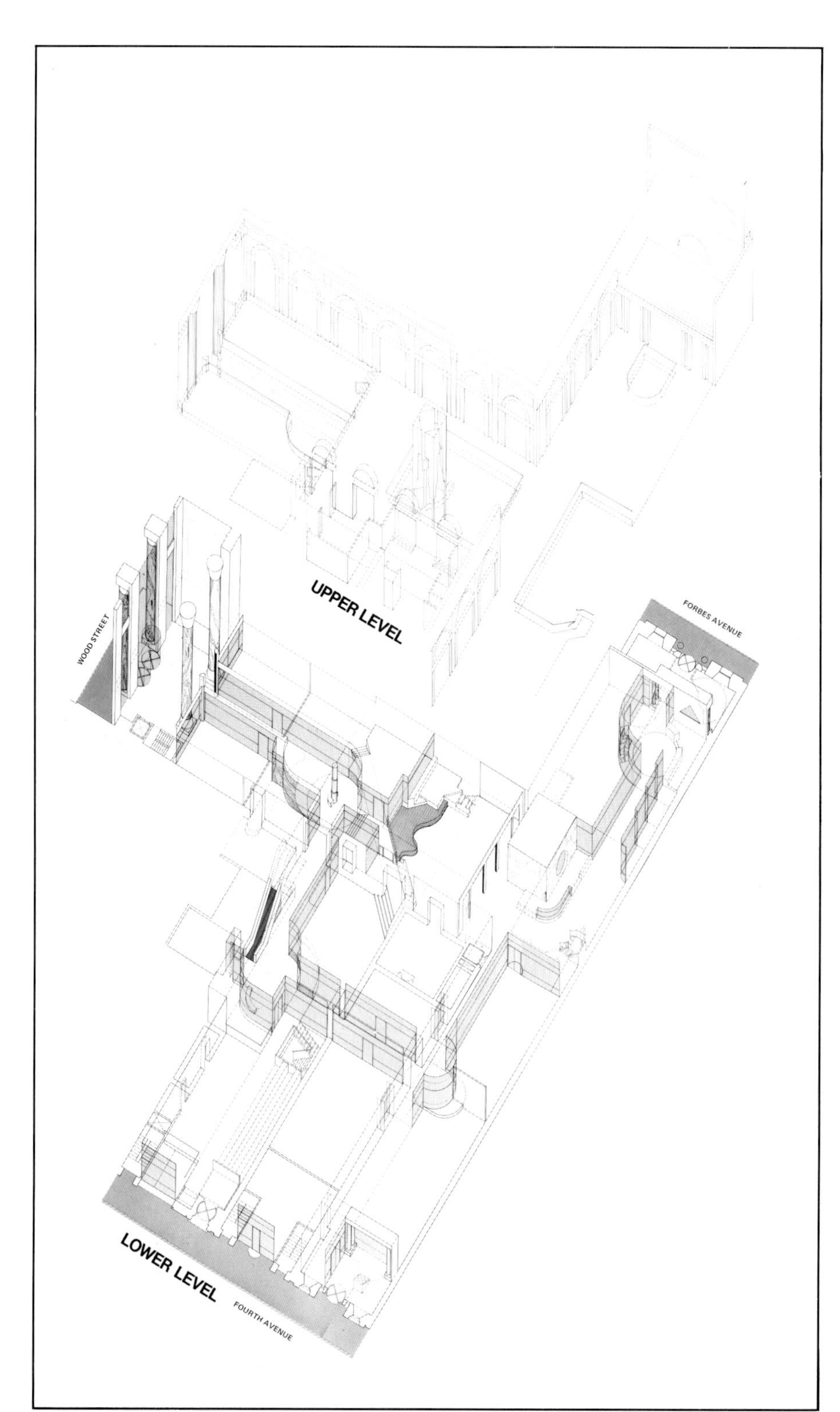

Interior *(Above)*. The original lavish interior shell has been retained as a background for a group of small shops and a cafe. The large decorative stained glass skylight and the rich marble columns, arches, and pilasters provide an attractive backdrop for the main atrium.

Isometric view *(Left)*. This bird's-eye-view shows how The Bank links entrances from three major thoroughfares—Wood Street, Forbes Avenue, and Fourth Avenue—into its central atrium, creating a complex and interesting space within the group of commercial buildings.

ONE CHURCH STREET

PROJECT
One Church Street
Nashville, Tennessee

ARCHITECT/DEVELOPER
The Ehrenkrantz Group, P.C.

STRUCTURAL ENGINEER
Ross-Bryan

CONTRACTOR
Hardaway Construction

Before restoration. Prior to restoration the nineteenth-century facade was grimy but intact, with the exception of the entrance storefront. The masonry portions of the facade were cleaned, the storefront restored, and all the wood trim painted. (The contrast is quite striking between the before and after photographs.)

Restored facade. This small facade in downtown Nashville gives no hint of the large modern office space that lies behind it. In a joint venture between the architects and local investors a neglected old warehouse has been adapted to attractive new commercial space.

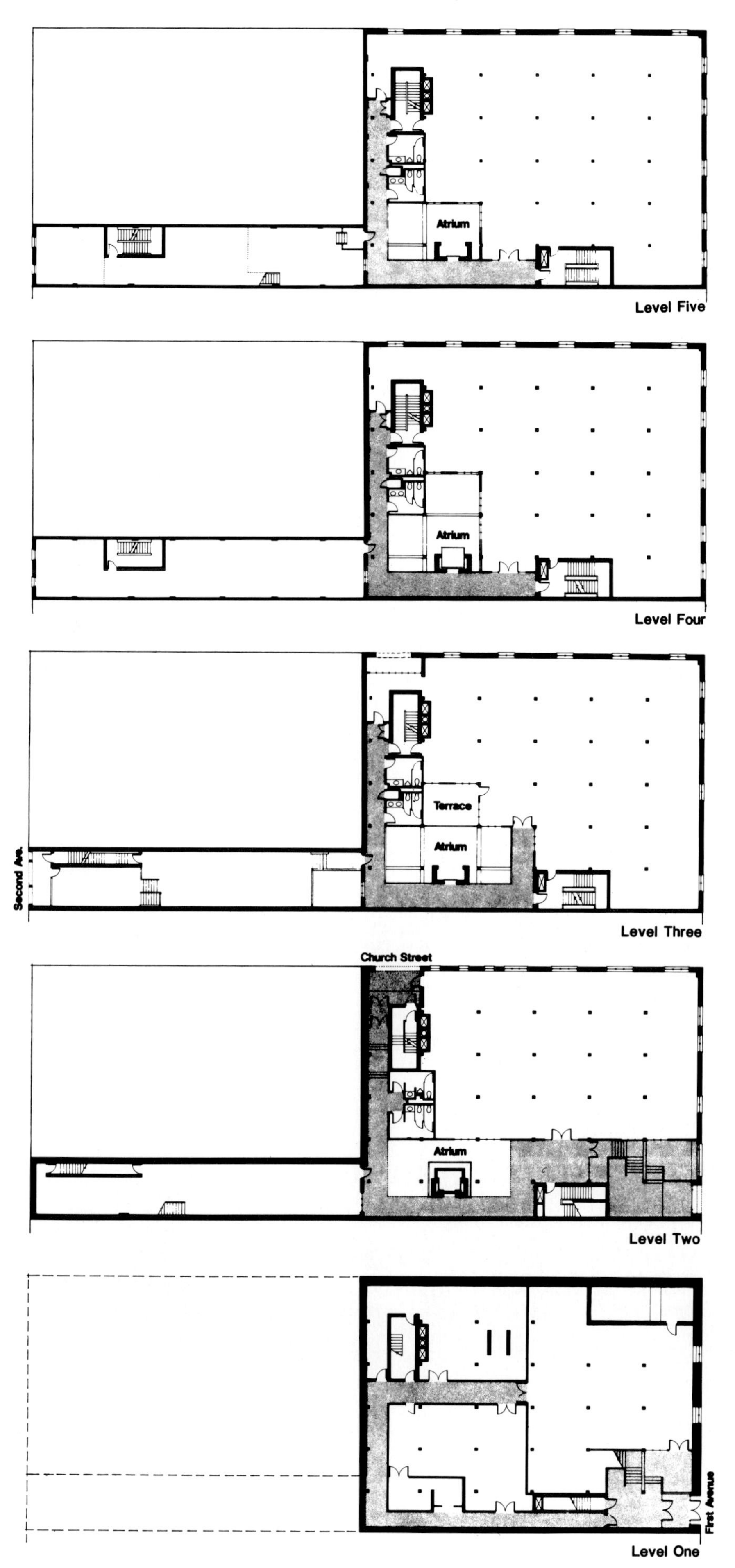

Plans. Studying the plans it becomes evident that this project is much larger and more complex than its diminutive nineteenth-century facade on Second Avenue indicates. The site slopes down hill, providing an additional story on First Avenue. The new main entrance from Church Street is at the cellar level of the smaller building. The new entry leads to a skylit atrium (containing a glass-enclosed elevator) that has been created to provide daylight in the windowless portion of the interior furthest from the surrounding streets.

Loft interior. A view of the old hoistway shaft in the former warehouse shows exposed brick walls, heavy mill timber framing, and wood joists typical of nineteenth-century commercial warehousing and manufacturing structures. (Note the exposed sprinkler pipes mounted on the ceiling.)

New atrium. The new entrance from Church Street leads to a formerly dark windowless interior of the warehouse, which is now flooded with daylight from the new skylit atrium containing a glass-enclosed elevator.

SEATTLE GARDEN CENTER

PROJECT
Seattle Garden Center
Seattle, Washington

OWNER
GCB Partnership

ARCHITECT
Arne Bystrom

ENGINEER
Darrold Bolton

CONSULTANT
Stan Volk, Smith Bros. Heating

GENERAL CONTRACTOR
Burfitt Construction Co.

RESTORING A POPULAR WATERFRONT MARKET

Existing Structure

A minor building formed an important visual element in restoring a picturesque waterfront market district in downtown Seattle, Washington. After World War II the traditional distribution pattern of wholesale produce markets began to shift from the waterfront and railroad to the vast new network of highways crisscrossing the country. Refrigeration and air transport further accelerated the decline of the formerly bustling urban harbors.

Ghiradelli Square, in San Francisco, was perhaps the first American development to recognize the potential for reviving our neglected downtown waterfronts for merchandising and tourism. Seattle's Pike Place Market is dramatically located on a steep hillside overlooking a magnificent harbor. The open-air produce market draws crowds of residents, while the food stalls and craftspeople's displays attract the tourists. These colorful surroundings provide an ideal setting for the small, flamboyant 1930s Art Deco style Seattle Garden Center Building.

Restoration Program

Architect Arne Bystrom bought the building together with two partners. On the street level opposite the Pike Place Market buildings, he provided a commercial space for each of his partners. On the upper level, the architect added a penthouse and rooftop terraces for his own office and pied-à-terre overlooking the bustling market, harbor ferries, and Puget Sound. By developing and adapting the structure to new uses and respecting the unique character of the neighborhood, he was able to provide himself with an attractive and inexpensive office.

Exterior Work

Between the time of the building's original construction and this renovation, West Coast architects and engineers have become much more knowledgeable about the seismic problems in their region. Located on a sloping hillside, the building contained a huge void that in places reached 12 feet in height. This cavernous hollow had to be filled with concrete, and steel knee-braces were added to shore up an existing brick retaining wall that supported an alleyway between the building and the upper hillside slope. The engineer redesigned the foundations to accept structural damage, but not failure, in the event of an earthquake.

Seattle Garden Center. The small one-story commercial structure steps up the hillside in the midst of a proposed urban renewal project in Seattle's Pike Place Market district.

The original exterior wall construction of concrete and brick was cleaned, repaired, and repainted. The interior is framed with 8 × 12-inch wood floor joists supported by 8 × 8-inch heavy timber posts. The new upper-story addition for the architect's office is of a similar construction, but to lighten the additional load on the foundations, architect Bystrom used fireproof studs in the exterior walls and used stucco-work to match the solid concrete of the walls below.

The architect found that the building had been painted in two bright pastel colors: salmon and green. The colors were faded, so when new paint was applied to replicate the original, the architect added a bright pink to highlight some of the geometric decorative details. This drew some criticism, but after a year or two the paint toned down and so did the controversy. This issue is not necessarily limited to modern buildings; researchers have discovered archaeological evidence that many eighteenth- and nineteenth-century color schemes were a great deal more vibrant than the canons of conventional good taste allow. Because color preferences are so subjective, it is very difficult to enforce this aspect of landmarks jurisdiction in commercial districts and especially in residential neighborhoods.

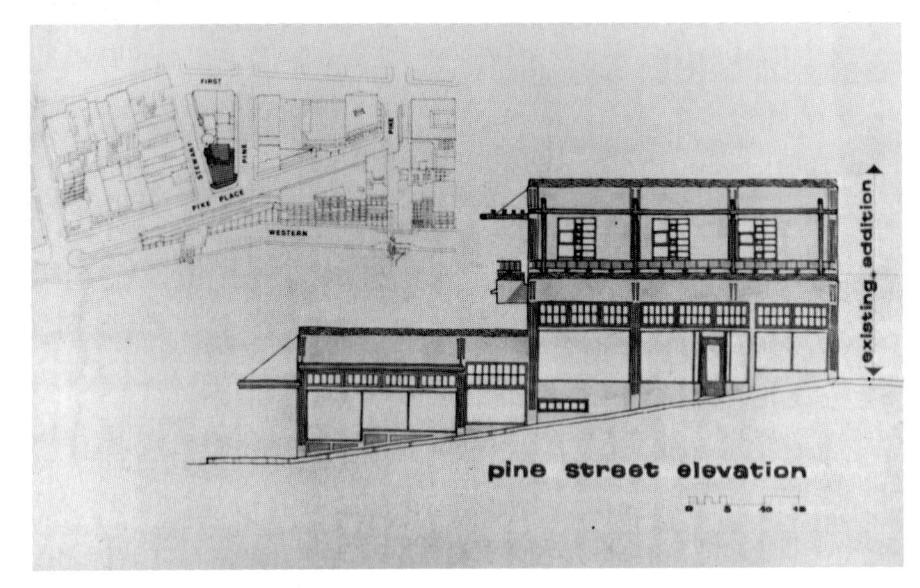

Site plan—elevation. The site plan indicates the Seattle Garden Center's key location in the midst of the Pike Place Market district. The Pine Street elevation indicates the additional story designed by architect Arne Bystrom for his own office.

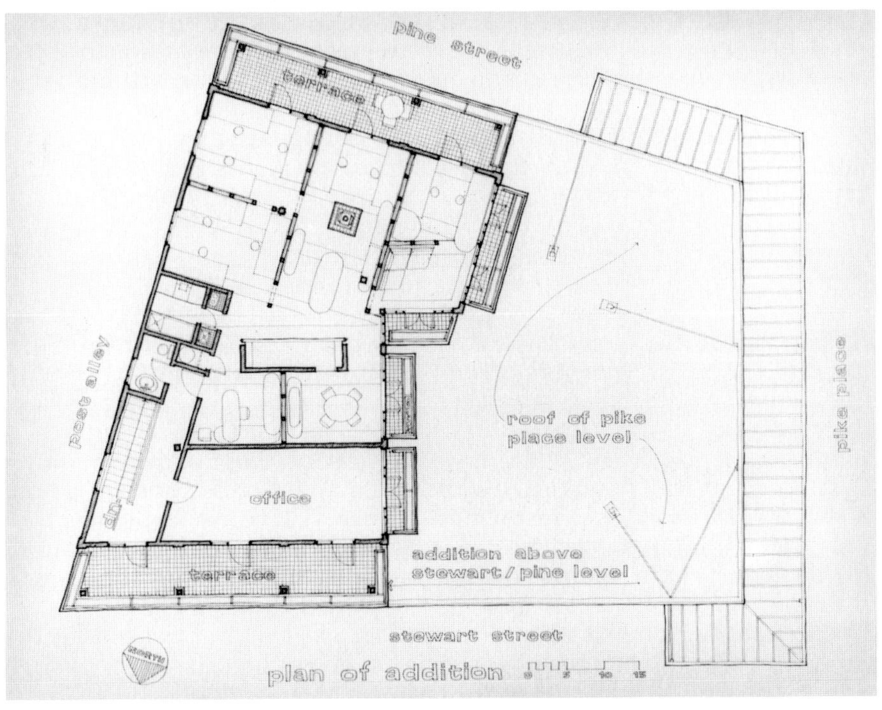

Plan of addition. The plan illustrates the compact layout that architect Arne Bystrom developed for his office within the irregular boundaries of the steep hillside site. The terraces open out to wonderful views of the harbor and Puget Sound.

Pike Place Market. On a steeply sloping site leading to the Pike Place Market an enterprising architect restored and enlarged a neglected small commercial structure originally slated for demolition. Its prominent location and colorful Art Deco detailing transformed it into a local landmark and provided the architect with a small office overlooking the harbor.

Interior

Whereas the commercial spaces were left as unfinished shells, the architect invented a unique interior for his penthouse office, combining the Art Deco spirit of the exterior with redwood millwork detailing reminiscent of Frank Lloyd Wright. One of the features that recycled loft and commercial structures offer is the opportunity to provide a stark contrast between old and new.

Project Finances

The city did not offer any financial assistance to the renovation project itself, which was the first in the Pike Place Market district to be developed without any grants. The partners financed the construction themselves and completed the building at $35 per square foot. Bystrom says that was relatively inexpensive by comparison to other Pike Place Market projects and to his least costly project, a warehouse space, constructed at $25 per square foot. The modest size of the building influenced the cost and neither elevators nor sprinklers were required. New electrical and plumbing systems were installed.

Project Aprovals

Initially, the Urban Renewal Agency had planned to demolish the building to make way for a larger project. Their decision also stemmed from the reports of several consultants that stated that the Seattle Garden Center Building was "hopelessly defective." The architect, Arne Bystrom, and the structural engineer, Darrold Bolton, both of whom had extensive experience restoring structures, were able to prove that it was, indeed, salvageable. The city of Seattle cooperated with the building owners, first by selling the building to them for only $40,000 (approximately $10 per square foot), and then by acquiescing in the matter of the color scheme. It also reconsidered its initial objections to the profusion of colorful hanging flower baskets and planters. (Now such natural decorations are taken for granted in the neighborhood.)

Comments
The Seattle Garden Center Building demonstrates that an imaginative architect with a good sense of timing can successfully become his own client. Many architects interested in recycling old buildings have become involved in design/build projects, contributing their design services instead of capital to the venture.

Office interior *(Above)*. Throughout this compact architectural office Arne Bystrom has meticulously detailed the native redwood millwork and built-in cabinetry in a manner reminiscent of Frank Lloyd Wright.

Office vista *(Left)*. Because of Seattle's sloping terrain this view from a small sitting area off the drafting room appears more like a penthouse in a skyscraper than the second story of a building in the middle of a produce market.

CANANDAIGUA NATIONAL BANK

PROJECT
Canandaigua National Bank
Canandaigua, New York

ARCHITECT
Mendel-Mesick-Cohen-Waite

STRUCTURAL ENGINEER
Klepper, Hahn and Hyatt

MECHANICAL AND
ELECTRICAL ENGINEERS
Robson and Woese, Inc.

Original building. An old photograph of the missing "gap" on Canandaigua's main street served as a guide to the architects designing an "infill" to link the adjacent structures.

Reconstructed infill structure. Using a historic photograph of the original structure as a guide, the architects designed a new infill structure to link the adjacent buildings and create a larger ground floor banking room. The reconstructed facade is not a literal facsimile of the nineteenth-century original.

Main Street, Canandaigua, New York. An obvious gap exists in the "street wall" of the turn-of-the-century commercial buildings lining the main street of a small New York State town. This problem is typical of many main streets in small towns all over the country.

Restored Main Street commercial block, Canandaigua, New York. In a typical nineteenth-century linear main street commercial block, a gap is far more noticeable than the materials or details of any particular architectural component of the ensemble.

PUBLIC BUILDINGS

All over the country, large and small communities are faced with the dilemma of aging and obsolete public buildings, courthouses, city halls, and so on. Many of these structures are architecturally and historically significant to their communities. Through careful analysis of their current condition, many of these public buildings can be made more functional and can be efficiently restored for extended use. It is important, not only economically but symbolically, to retain the dignity and integrity of traditional landmark buildings and institutions if one wishes to encourage the private sector to support historic preservation

CHENANGO COUNTY COURTHOUSE

PROJECT
Chenango County Courthouse
Norwich, New York

ARCHITECT
Mendel-Mesick-Cohen-Waite

STRUCTURAL ENGINEER
Eckerlin, Klepper, Hahn & Hyatt

MECHANICAL ENGINEER
Robert D. Krouner

LANDMARK COUNTY COURTHOUSE RESTORED FOR EXTENDED USE

Existing Conditions

The citizens of Chenango County were faced with a predicament not uncommon in many established rural communities: What do you do about the old county courthouse? The challenge was finding a way to prevent demolishing and replacing the landmark 1830s Chenango County Courthouse, which had become obsolete for contemporary needs. Local officials were tempted to solve the problem by demolishing the old courthouse and replacing it with an extension built on to an existing county office building.

The courthouse faces the town green in Norwich, a town of 10,000 inhabitants in a rural farming region near Cooperstown in central New York State. Time, tourism, and large-scale development seem to have passed this community by. The Greek Revival courthouse is a country cousin of its more high-style architectural kin along the Eastern seaboard, with its rustic proportions no doubt copied by local builders using one of the popular carpenters' manuals of the period.

Restoration Program

The decision to retain the structure was feasible because the difficulties were functional in nature and a great increase in space was not required. By the time Mendel-Mesick-Cohen-Waite was called in 1977, the local debate had been going on for 20 years. The architects' first step was to research and prepare a historic structures report to document the history and original features of the courthouse as well as determine the subsequent modifications to the restoration. They then developed a proposal for upgrading the courthouse's physical structure, functional layout, and mechanical systems. Once this was complete, they could demonstrate that the cost of restoration and rehabilitation would not exceed the cost of demolition and replacement with a new building. The architects also assisted in organizing a public campaign for private contributions to restore the old courthouse.

Project Finances

A committee of county legislators, judges, and private citizens was organized to explore the restoration of the courthouse. The restoration committee retained the architectural firm of Mendel-Mesick-Cohen-Waite to make a historic structures report partly funded by a grant from the National Historic Preservation Act. The cost of restoration and construction was aided by funds from the U.S. Department of the Interior (through the New York State Historic Preservation Officer) and from the local public works grant program of the Economic Development Administration of the U.S. Department of Commerce.

Chenango County raised the larger part of the funding and tapped into federal revenue-sharing funds. Some private donations were received. When the restoration was completed in 1980, the cost of $84 per square foot was comparable to the cost of a new building. However, by preserving the old courthouse, Chenango now has a building of traditional symbolic stature, richness, and dignity that could not have been obtained in a new building at any price.

Exterior

Most of the modifications to the courthouse during the century and a half since its construction were on the inside, so that restoring the exterior was fairly straightforward. Even though the original windows were 12-over-12, this was inconsistent with high-style Greek Revival; at a later date, they had been replaced by vertically divided sash. The original 12-over-12 windows have been restored, but the new sash has been double-glazed to conserve energy.

One of the most distinctive features of the old courthouse was a statue of Justice atop the cupola. Originally carved in wood by craftsmen who provided figureheads for ships, it has been replaced with a more weather-resistant replica of reinforced polyester. Four old chimneys, which were once part of the rooftop silhouette, were replicated and integrated into a ventilation system that also incorporated the existing nineteenth-century sheet metal rooftop ventilator.

Exterior. The exterior of the restored Chenango County Courthouse retains its traditional dignified presence on the Norwich, New York, town green and gives no hint that it has been upgraded and refurbished on the interior to cope with the contemporary functional requirements.

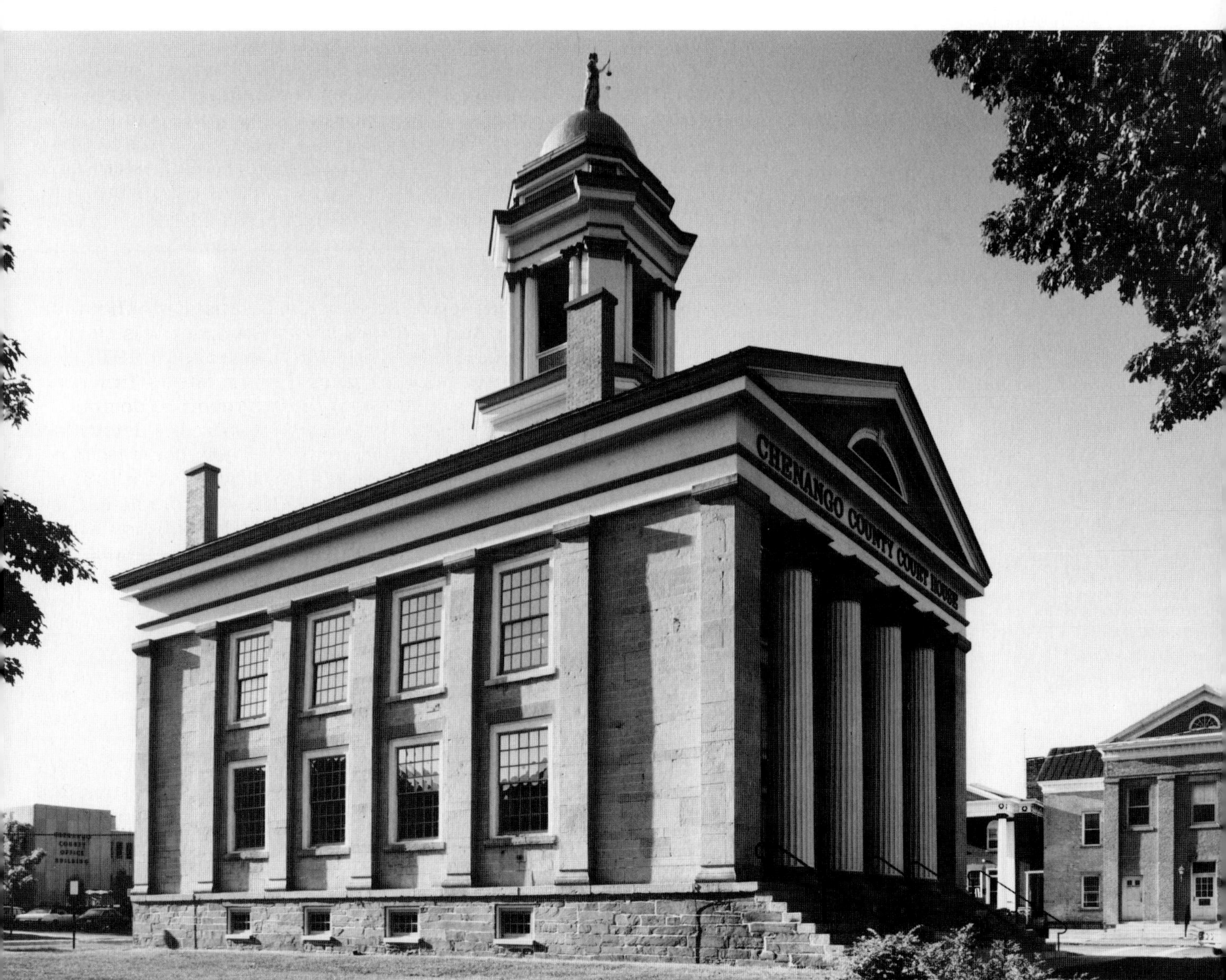

Interior

Because of the many changes to the building during its 150-year history, the architect took an eclectic approach to the restoration, incorporating some later features and removing others. Because the upper story courtroom itself had been extensively remodeled at the turn of the century, it was decided to restore this interior rather than the original.

Suspended ceilings, which in some areas lopped off as much as 4 feet of the tall windows, were removed. The old ornate pressed-tin ceilings were refurbished; missing tin panels were replicated in polyester. Sometimes the ceilings were lowered to improve the acoustics and to reduce the volume of heating space. Period carpeting, double glazing, and a new mechanical system were provided in the courthouse restoration.

Another contemporary material, vinyl tile, was chosen to replace the original oilcloth floor covering. The original oilcloth had been covered with successive layers of later floor coverings, and the architects considered replacing it with wood or marble. Vinyl was finally selected because it could form a checkerboard pattern and it possesses textural and acoustical qualities similar to oilcloth. Also, vinyl and oilcloth are both "common" materials that will wear out and have to be replaced in the future. It is often difficult to resist the temptation to replace more humble materials with more permanent and serviceable ones. The original color schemes were restored after careful scraping through layers of paint applied over the years.

Because the main staircase had insufficient headroom, the architect had it removed and within the same space designed a curved stair with adequate headroom. A second staircase, to serve as an emergency exit, was built at the other end of the building, and an elevator was installed to provide access for disabled people.

A new electrical system had to be installed. The old courtroom and principal corridors are now lighted with reproductions of the late nineteenth-century brass gasoliers. Conventional fluorescent fixtures illuminate the offices and behind-the-scenes work areas.

Modern heating and air conditioning were easily recessed into the existing chases within the thick masonry walls. They are served by a two-pipe fan coil system. To conserve energy, new insulation was installed in the attic spaces and the basement walls, and double glazing was used for the windows.

Original interior *(Above)*. An old photograph was helpful to the architects in restoring the turn-of-the-century character of the courtroom. (Note that the light fixtures appear to have been located along the central aisle and would not have provided a very good or even distribution of light.)

Prior to restoration *(Right)*. This photograph shows the courtroom prior to the restoration. (Note that the original coffered pressed-tin ceiling has been concealed by a modern suspended acoustical surface, stem-mounted fluorescent fixtures have replaced the earlier gasoliers, and the earlier fixed auditorium seating has been replaced with wooden pews.)

After restoration. The architects have taken an eclectic approach to the project, restoring some features and substituting others. More electrified reproductions of period gasoliers have been placed in the courtroom than were there originally to provide better illumination. The tin ceiling has been restored and wall-to-wall carpet installed to improve acoustics and reduce maintenance.

Comments

Public buildings require more than an "authentic" restoration. They must conform to current standards of safety, security, and convenience. The courthouse now meets those standards, and through careful design and attention to detail, these modifications have been tucked away unobtrusively.

Detail of restored pressed tin ceiling. Removal of a suspended acoustical ceiling and fluorescent light fixtures, typical of post World War II remodelings, revealed the earlier decorative pressed tin coffered ceiling treatment which was intact.

CHAPTER SIX

EVALUATING THE STRUCTURE

INSPECTION, NEW-USE REQUIREMENTS

In order to undertake the preservation of old buildings, it is essential to be familiar with traditional methods of constuction. Costly delays and errors can be avoided if you recognize inherent limitations hidden behind old walls and finishes and plan accordingly. Most old and new structures follow certain vernacular and regional patterns of construction and detailing. With some knowledge and a little experience you will know what to expect.

INSPECTION

The first step in restoring an old building is to examine its physical condition along with the materials and methods that were employed in its construction. Once this has been done, you can determine how much repair, restoration, or replacement is required and you can then evaluate the most appropriate materials and methods to accomplish your objectives.

Examining a building to determine its physical condition can be revealing if you know where to look and what to look for. Since most buildings are assemblages of many different, interacting materials, it may take some detective work to uncover the cause of problems. Signs of deterioration can be misleading; for example, a damaged and sagging

Figure 6-1 Matteson Tavern, Shaftsbury, Vermont. After a historic structure suffers serious fire damage a temporary program of stabilization must be undertaken until a full assessment can be made. Typically, this would include sealing the structure against weather exposure, sealing up openings to prevent vandalism, and shoring to prop up weakened structural elements.

Figure 6-2 Cracked sandstone column, former Astor Library, New York, New York. It is usually less complicated to repair severely deteriorated masonry facade elements in situ, since their removal might require shoring. A temporary method of securing loose or cracked masonry elements should be used until the repair can be made. Fragile carvings, cornices, and projecting elements must also be protected against accidental damage from scaffolding during routine building cleaning and facade repair work.

plaster ceiling may require more than a plaster repair. It usually indicates that a leak somewhere has caused a loss of bond to the lathing material. Tracing the source of a leak may reveal an interior plumbing problem or, perhaps more seriously, a deteriorated roof that must be repaired before the plaster can be restored.

Some old buildings are too heavily damaged to be salvageable (Figure 6-1). One of the first efforts in saving an old building should be installing a temporary stabilization program to halt deterioration until a viable restoration program can be organized (Figure 6-2). This may mean getting an emergency contract to install a temporary roof of tar paper, vinyl sheeting, or corrugated plastic over plywood. It may also be necessary to install temporary shoring, bracing, or buttresses until the extent of structural damage can be assessed. In extreme cases it may be determined that only the exterior masonry shell can be salvaged and all the wood interior and roof framing must be replaced (Figure 6-3).

Exterior Inspection

When making an exterior inspection, try to find evidence of defects or deterioration. Flaws will vary depending on the type of construction and the materials used. Once you have identified the problem areas, you can begin to investigate the probable causes and determine the cost of solving the problem.

In masonry buildings, cracks, loose brick or stone, open joints, and staining are signs of potential problems (see Figure 6-4). In wood buildings, warped, rotted, or missing elements should be examined carefully. In buildings constructed of several materials, such as steel-framed structures faced with brick, terra cotta, or stone, finding the source of the deterioration can be extremely complex and frustrating. In tall buildings, look for signs of cracking and deterioration due to moisture penetration caused by weather exposure.

Figure 6-3 Cracked finial, former Astor Library, New York, New York. A severely cracked sandstone decorative urn finial on top of the main facade masonry parapet has been temporarily wrapped in metal mesh until repairs are made.

Figure 6-4 Nineteenth-century brick and brownstone row house, Brooklyn, New York. A conspicuous diagonal crack running through the brick wall and brownstone water table observed during an exterior inspection merits further investigation.

Also, examine the condition of the exterior walls where the building walls meet the ground. If you find indications of deterioration in the masonry foundations or dry rot at the wood sills, check to see if there is proper site drainage. Dampness in a cellar can be more than mildew; it may be a sign of serious foundation problems. The other most likely source of water penetration is the roof.

Roofs. Wisely, folklore is full of proverbs about keeping a sound roof overhead. The roof is indeed an important place to start an inspection, because water penetration at the roof is usually the most serious cause of deterioration in older buildings. If neglected too long, there can be irreversible damage to a building's structural integrity. Many abandoned and vandalized old structures, open to the weather for long periods, are so compromised that a rescue effort is economically not feasible.

The evidence of roofing failure usually shows up inside before it is visible on the outside. (See Figure 6-5.) At first, damp patches will appear after heavy rains or snow, and if nothing is done, water will eventually drip from the ceilings and stain the walls. Sometimes, moisture penetration through the wall is simply a matter of poor drainage, blocked gutters, or backed-up leaders—all of which can be taken care of through routine maintenance. Roof leaks sometimes penetrate through walls and ceilings to lower floors, but more often interior water damage on lower stories can be traced to deteriorated plumbing.

If the flashing or the roofing material has failed, tracing the source of water penetration can be much more difficult. Once the probable cause of the leak has been located, the roof framing in the vicinity must be allowed to thoroughly dry out before being sealed with new waterproofing. If moisture becomes trapped in insufficiently ventilated roof spaces, wood framing members may begin to rot. It is not uncom-

Figure 6-5 Nineteenth-century Queen Anne roofs, Brooklyn, New York. Note the evident patching and replacement of original copper flashing and slate with asphalt shingles and roll roofing. Even a superficial investigation would probably reveal water penetration and damage on the interior.

Figure 6-6 Lafayette Street Firehouse, New York, New York, 1895 (Napoleon LeBrun, Arch.). The New York City Fire Department was forced to abandon a landmark Chateausesque firehouse built on pilings of an eighteenth-century reservoir. Later, office building construction and a nearby subway lowered the water table exposing the pile caps and caused settlement that loosened some limestone facade elements.

mon in inner-city row-house neighborhoods to replace most of a building's roof and floor joists even if the rest of the building is in good physical condition. Even when minor roofing repairs are planned, the replacement work must be carefully phased so that the building is not exposed to damage from sudden storms. If the replacement work must be prolonged or complicated, as often happens with roofs of slate or mission tile, it may be necessary to construct a temporary scaffolding, shed, or roof over the exposed portions of the structure.

Foundations. Once the integrity of the roof and roof drainage systems (flashing, gutters, leaders, and so on) has been checked, the exterior walls and the foundations must be scrutinized from top to bottom.

Before 1900, most buildings were constructed on spread footings or rubble walls built on a bed of flat stones laid directly on the earth. Changing levels of ground water, rain runoff, and other subsoil conditions can and did cause uneven settlement of these rubble foundations. Check the cellar for evidence of dampness, cracks in foundation walls, or settling. If evidence of extreme dampness is found, it may be necessary to expose the foundation walls, damp-proof them with a coating of an asphalt compound, and install a piped foundation drainage system. Of course, the outlets at the ground of all roof leaders should be checked to make sure that rain runoff is being discharged far enough away from the house to avoid damage to the foundations.

In the nineteenth century, masonry buildings were sometimes constructed over wood pilings on landfills, coastal and swampy areas (see Figure 6-6). Deteriorated pile caps can seriously undermine the structural integrity of an old building. Underpinning and replacing these deteriorated rubble foundations is tedious and costly.

In urban areas, new excavation and blasting for the construction of multistory structures, underground utilities, and tunnels adjacent to old buildings can disturb and undermine existing foundations, causing structural damage. Prior to any alteration, all vertical cracks running through the exterior masonry should be investigated. Even if they are determined to be of no significance, should be monitored during construction. Cracked door and window sills and lintels need to be stabilized or replaced. A structural engineer with special experience in foundations should be engaged during major excavation work.

Interior Inspection

An interior inspection should be related to the problems of the exterior, since often the telltale evidence of water penetration may show up as a benign-looking damp spot on a wall or mildew in a closet. It is important to distinguish between water damage caused by a problem with the exterior shell or roof and that which results from bad plumbing.

After checking for leaks, the next important concern is structural stability. Sloping floors and ceilings are endemic to old buildings and are not necessarily cause for alarm. However, excessively springy floors, balconies, or stairways may indicate that the structure is overloaded or deteriorated in some way. Some deflection may be corrected by jacking up the old joists until they are level (see Figures 6-7 and 6-8). But excessive deformations cannot be corrected. The case study in Chapter 5, on the Whitefield Mansion in Southampton, New York, includes a description of a novel method of leveling a sagging floor.

Figure 6-7 Replacing floor joists, Fort Herkimer Church, New York. The rotted wood floor framing in an eighteenth-century church are being replaced. Generally the wood structural members of an old building are more prone to deterioration than masonry elements. Frequently, dry rot occurs where wood is located close to the earth.

Although most floor framing members are concealed by plaster ceiling finishes, the floor over the cellar is usually exposed. A quick visit to the cellar will show whether dry rot has attacked the beams or joists and if props or posts are shimming up the sagging floors. Sometimes the placing of additional vertical support is a legitimate method of increasing floor loading by shortening the span of old or undersized framing members.

ASSESSING FOR NEW-USE REQUIREMENTS

In considering any adaptation to new uses, it is essential to investigate the physical condition and load-bearing capacity of an existing old framing system. (See Checklist 6-1.)

Load-Bearing Capacity

The original residential loading capacity of a old row house may be far short of that needed for its new use as a commercial office typically furnished with heavy steel filing cabinets, desks, business equipment, and computers. In contrast, an old loft manufacturing building, constructed of heavy timber framing, will probably have a strong enough floor-loading capacity for the new commercial uses, since it was originally designed for warehousing or heavy machinery. Although estimating loading capacity of timber framing is not simple, it is especially difficult to determine the effective loading capacity of late nineteenth-century cast-iron framing and hollow terra cotta tile vault systems. The only practical and reliable option is to have a historical engineer or a civil engineer familiar with old structural systems supervise a load test.

Figure 6-8 Shoring a column, Fort Herkimer Church, New York. In order to make required repairs to deteriorated wood framing elements it is sometimes necessary to provide temporary shoring. Wood load-bearing columns are being temporarily supported while the rotted flooring joists are replaced.

Once the loading capacity has been established, measures can be taken to bolster the loading capacity if modern building code provisions require it. Even if the old framing is structurally adequate, most building codes require some form of fireproofing, often including the installation of automatic systems. Preservation efforts to keep the original architectural detailing and appearance are very often at odds with code enforcement officials' concerns about safety.

Strengthening or bolstering the original framing system to accommodate loading requirements is easily accomplished if the space to be converted is located immediately over an accessible space, such as a basement or unfinished cellar. However, if the upper stories are to be used and they have intact old decorative plaster ceilings attached to the original framing, the task is much more complex.

Checklist 6-1
INSPECTING AN OLD BUILDING

EXTERIOR

Roof

- ☐ Is metal flashing at joints and intersections loose or damaged?
- ☐ Is gutter system corroded, set at incorrect pitch, or undersized?
- ☐ Is ridge sagging?
- ☐ Is roofing material itself in good condition?
- ☐ Are tiles or shingles loose, missing, cracked, or worn?

Foundation

- ☐ Are foundation walls and sills cracked?
- ☐ Does masonry or wood show signs of excessive dampness?

Walls

- ☐ Does masonry have loose bricks or stones, vertical cracks, open joints, or rust stains?
- ☐ Is brownstone or cast concrete crumbling or eroded?
- ☐ Is wood warped or rotted?
- ☐ Is paint chalking, blistered, peeling, or cracked?

Doors and Windows

- ☐ Are sills, lintels, and sashes in good condition?
- ☐ Does glass require replacement?
- ☐ Does size and location of doors and windows provide adequate exit and ventilation?

Decorative Elements

- ☐ Is cast-iron facade rusted, corroded, or cracked?
- ☐ Are wood elements, such as shutters and porch railings, rotted or missing?
- ☐ Are terra cotta or stone ornaments loose, eroded, or stained?

INTERIOR

Attic

- ☐ Are rafters and ceiling joists drooping or rotting?
- ☐ Is sheathing deteriorated?

Cellar

- ☐ Are floor joists and beams in good condition?
- ☐ Are props or posts shimming up sagging floors?
- ☐ Are walls cracked or excessively damp?

Floors and Stairways

- ☐ Are floors or stairways excessively sloped or springy?
- ☐ Is wood warped or decayed?
- ☐ Is load-bearing capacity sufficient for intended use?

Ceilings and Walls

- ☐ Are there bulges or cracks in the plaster?
- ☐ Are decorative moldings and medallions broken or missing?

Plumbing and Mechanical Systems

- ☐ Are pipes for water supply and waste disposal in good condition?
- ☐ Do vents and traps on waste lines meet present code requirements?
- ☐ Will electrical wiring, heating, ventilation, and fire-protection systems have to be upgraded?

Upgrading Mechanical Systems

Usually, in addition to increasing the floor-loading capacity, codes require the upgrading of old systems with modern electrical wiring, plumbing, fire protection, and mechanical ventilation. These are safety measures with major economic and social implications, and neglecting code requirements may result in higher insurance premiums. In most projects requiring massive installation and concealment of new code-required systems, the original ceiling finishes must be replaced. In many cases, so much of the original interior detailing is already missing, or is in such poor condition, that reproduction and replacement is the only practical course. Extensive changes in layout and the addition or relocation of door and window openings may require the reproduction of existing wood and plaster architectural detailing to achieve a consistent effect.

EXTERIOR CLEANING

Many old buildings whose exteriors are faced with traditional materials such as brownstone, glazed terra cotta, and cast iron have not weathered well and may require extensive repairs or replacement of seriously deteriorated components. Acid rain and environmental pollutants have greatly accelerated the deterioration of old buildings, outdoor statuary, and bridges. A distinction must be made between cosmetic cleaning to remove grime and to improve an old building's appearance and restoration that is done to arrest decay and deterioration.

Cleaning and restoration require expert knowledge of the long-term effects of methods and materials and their interaction with atmospheric conditions. Unless properly executed, cleaning can cause serious damage. Cleaning is also complicated by materials used in the construction of old buildings. An effective cleaning process for brick may be extremely injurious to brownstone or terra cotta elements within the same facade. Masking or protecting these elements may be impractical. (Various cleaning methods are discussed in Chapter 7.)

MOVING OLD BUILDINGS: PRO AND CON

Exposure to the elements accelerates decay in old buildings. Also, the physical deterioration and neglect of the surrounding area may result in the abandonment of a structure that might be treasured in another neighborhood.

Technology today permits incredible undertakings that previous generations could only fantasize about. There is nothing new about moving buildings, but in some respects the process is more complex today because of the mechanical systems linking an old building to its site. Water supply, gas and sewer connections, and electrical communications are all lost in even the shortest move. Preparation costs, the move itself, and the re-creation of a new foundation and new site utilities make moving a costly undertaking.

In addition, there are other issues to consider before deciding to move a historic structure. Perhaps the most negative aspect of moving an old building is severing its sense of continuity with its original site. Moving an old building does requires replacing more of its historic fabric than a restoration in situ because some elements are too fragile to survive a move. There are circumstances, such as profound deterioration of the original surrounding environment, that leave no reasonable alternative to relocation. However, the extreme to which relocation can go is being seen today in a nostalgia industry that caters to old-building buffs (see Figure 6-9). For example, charming, genuine, old farmhouses from unspoiled but economically depressed areas of New England are appearing, seemingly overnight, at sites like Long Island's fashionable Hamptons.

Figure 6-9 Old Victorian, new setting. To rescue surviving landmarks stranded between the modern freeways, San Diego has relocated them to an artificial "Old Town." This "new" arrival looks strangely out of context.

Disassembling an Old Building

When an old building is being relocated beyond a reasonable radius of its original site, it usually becomes impractical to move it intact. Nothing provides a better object lesson in traditional methods of construction than the dismantling of an old building.

Old buildings do not come with a set of illustrated assembly instructions. Providing diagrams for reassembly as well as labeling all the parts is a slow, complicated, and painstaking process. Both disassembly and reconstruction require skilled craftsmanship and a high degree of supervision (Figure 6-10). The cost of this process can be staggering and is difficult to justify unless the building is of exceptional quality. Nonetheless, this is the way William Randolph Hearst assembled his European treasures to create San Simeon.

Figure 6-10 Independence Plaza, New York, New York. To preserve a few fragile landmark structures in the Washington Market Urban Renewal Project they were moved to create the impression of a row. Only the exterior shells were restored and they were auctioned by the city with unfinished interiors.

CHAPTER SEVEN

KNOWING THE MATERIALS

IDENTIFICATION, REPAIR, RESTORATION

In a general survey, it is impossible to cover all the details of traditional materials and methods of construction. But it is important to understand how and of what old buildings were originally constructed, so that we can choose appropriate techniques for preserving them and, if necessary, can reproduce deteriorated or missing architectural elements. Another reason to understand traditional construction methods is to be able to anticipate and satisfy modern functional requirements—wiring, plumbing, and mechanical systems—in the least intrusive manner.

ORIGINAL MATERIALS VERSUS REPLACEMENTS

In many cases, utilizing the original materials and methods is still the most effective course, although modern materials and construction techniques should not be categorically ruled out (Figure 7-1). Sometimes the issue is more philosophical than practical. In an eighteenth-century museum restoration project, for example, it may be justified to replace crumbling and cracked plaster partitions with hand-split wood lath, but in most cases applying new plaster over wire mesh simulates the same effect much less expensively. When it is a matter of reproducing hand-carved wood or stone detailing, there may not be any alternatives to the "real thing."

Figure 7-1 Roof and facade restoration, City Hall, Oswego, New York. Crated reproductions of cut stone facade and coping elements awaiting installation at the job site. Even though the majority of the stonework is in good condition, certain portions of copings, projecting details, cornices, balconies and freestanding elements, such as balustrades, are especially prone to deterioration from weather exposure.

Figure 7-2 Chipped brownstone, former Astor Library, New York, New York. Brownstone cut into sharp profiles and corners is prone to chipping and spalling. In addition to being easily damaged by impact at street level, it crumbles as a result of spray from snow-melting salts and compounds.

Figure 7-3 Deteriorated carved brownstone architectural detail, former Astor Library, New York, New York. Brownstone (sandstone) is a soft stone which is easily carved. Unfortunately it deteriorates quickly due to weather exposure. Architectural ornaments sculpted in high relief are particularly vulnerable.

Figure 7-4 Deteriorated terra cotta, Woolworth Building, New York, New York. The condition of this decorative spandrel ornamented with Gothic motifs was typical of the random pattern of deterioration found all over the exterior of the Woolworth Building prior to its massive repair and restoration program.

Old is not necessarily better, and we are often obliged to improve on original materials and architectural detailing that have not withstood the test of time. For example, the nineteenth-century passion for brownstone and the early twentieth-century enthusiasm for glazed terra cotta are not shared by current owners whose investments face nagging and costly maintenance problems (Figures 7-2 and 7-3). Owners of skyscrapers with early terra cotta facings, such as the Woolworth Building (see Figure 7-4) in New York City and the Wrigley Building in Chicago, have spent millions of dollars on facade restoration. Both of these major projects experimented successfully with fiberglass and polymer concrete to replicate damaged or missing original terra cotta elements.

MASONRY

Masonry encompasses many materials, including brick, fieldstone, limestone, sandstone, marble, granite, and terra cotta. Sometimes several of them are combined, with brick usually serving as the load-bearing construction material and one or more of the others used as facing or decorative elements on the exterior walls. Masonry construction can endure for centuries, but when several types of masonry are combined the problems of restoration are more complicated (Figure 7-5).

Fieldstone

Fieldstone, or rubble wall, construction was prevalent during the colonial period and later during westward expansion when land was being readied for farming. The pattern of usage varies regionally, depending on the availability and character of the native stone. In New England there was so much stone strewn about the landscape that even the walls enclosing farm fields were constructed of "dry," or mortarless, fieldstone rubble. Often the foundations and cellars of farmhouses, barns, and outbuildings were constructed without mortar to allow natural site drainage and water run-off. The site drainage systems of these rustic buildings must be reworked when adapting them for residential use, or their new owners will be plagued by leaky cellars.

Until the early twentieth century, when continuous reinforced concrete foundations were developed, most structures, including urban row houses, were built on spread footings of flat slates or stones; on top of which were constructed rubblestone walls to the height of the basement story. The brick walls of the upper stories were erected on this stone base. Because of the compaction of landfill and uneven subsoil conditions, over a long period of time, differential settlement occurred. Old foundations can be stabilized, but underpinning is a tedious and costly process.

It is sometimes necessary to excavate within the existing foundations to increase minimal headroom in the cellar or basement. If this is done a section at a time, new, continuous concrete footing can be created below the rubble walls. Serious consideration must be given to any enlargement, addition, or major new construction adjacent to an old building because of the risk of undermining the existing foundations. Particularly if the new excavations will be deeper, it may be necessary to reinforce or buttress an old building prior to digging.

Skilled masons of the past, using carefully selected fieldstone to avoid the need for cutting, could not always avoid irregular mortar joints that were prone to moisture infiltration. In portions of New Jersey, Pennsylvania, and Delaware where the native stone is a porous variety of sandstone, it was customary to render the exterior walls with a whitewashed cement plaster. It is quite difficult to trace the source of leaks in this kind of rubble wall.

In the southern coastal regions of Georgia, a pumicelike stone containing seashell fragments is used in a rubble-adobe wall construction technique called "tabby." In other regions, local stonemasons displayed their versatility by employing anything from small polished cobble-

stones to giant boulders. The earliest fieldstone structures incorporated some brick, even though it was costly and often not locally produced, to create uniform window and door openings and chimney flues.

Figure 7-5 Restoration of former Astor Library facade, New York, New York. The deterioration of the brownstone architectural elements on this 1849 facade imperils the integrity of this landmark structure. A massive restoration program was organized by architects Mendel-Mesick-Waite-Cohen to stabilize the brownstone and restore the crumbling architectural detailing.

Cut Stone

Until the Industrial Revolution provided cutting machines and cheaper transportation from quarries, building with stone and marble was prohibitively expensive. Along the New England coast where granite is abundant, it is not unusual to find modest wood dwellings with massive cut stone steps and foundations. During the nineteenth century when cut stone became more available, it was frequently used as a veneer applied to brick bearing walls and was used structurally for columns, window heads, sills, steps, and so on.

Brownstone reached the height of its popularity in the so-called Brown Decades from 1850 until 1900 (see Figures 7-6, 7-7, and 7-8). Actually sandstone, brownstone is extremely soft and easy to carve, but it is also porous and vulnerable to freezing and thawing cycles. Brownstone was almost always used as a facing material, but because of heavy demand, the quality of quarrying and setting methods were not always first rate. Now, a century later, increased atmospheric pollution has reduced many brownstone facades to a disastrous state (Figure 7-9). Patented stabilizing and patching compounds have been promoted over the years but few have demonstrated any long-term success.

Erosion, staining, and deterioration may affect only certain portions of decorative stonework, depending on the degree of exposure to the weather and such features as sheltering cornices or overhangs. Discoloration may occur from rusting metal architectural elements such as bronze flagpole mounts, iron balconies, and window air conditioning units. Matching stonework is often difficult because many old quarries are no longer functioning. Even if the original source is still available, it is almost impossible to match the patina of weathered stone. It is usually easier to achieve a good match in a newly cleaned building.

Figure 7-6 Brownstone, the Church of St. Ann's and The Holy Trinity, Brooklyn, New York. The unsheltered molded or carved architectural elements such as buttresses, copings, and balconies are particularly prone to deterioration. The exterior of this nineteenth-century Gothic Revival church required a massive repair program.

Figure 7-7 Brownstone repair, the Church of St. Ann's and The Holy Trinity, Brooklyn, New York. After chopping away the loose, flaking, and deteriorated portion of the brownstone, the surface is roughened or scored so that the cement patching material will adhere.

Figure 7-8 Brownstone, the Church of St. Ann's and The Holy Trinity, Brooklyn, New York. To accurately restore damaged molded details, a template is taken of the existing profile prior to chopping off loose surface elements. The new molded profile is troweled using the template as a gauge.

Figure 7-9 Deteriorated brickwork, Federal town house, New York, New York. Repointing is necessary to restore the structural integrity and improve the appearance of this venerable structure which still retains its rare, original decorative window lintels.

Figure 7-10 Facade restoration, Old St. Ann's Church at Packer Collegiate Institute, Brooklyn, New York. Restoring a High Victorian Gothic landmark, such as this former church (now a private school), is complicated because of the variety of materials used in its construction. Scaffolding is frequently required in order to thoroughly determine the condition of inaccessible portions of the structure.

Brick

During the colonial period, imported brick was very quickly superseded by domestically produced brick (Figure 7-10). In the seventeenth and eighteenth centuries, brick was molded by hand and fired in kilns. Sizes were not standardized but generally followed European precedents. In the seventeenth century, the Dutch produced a yellowish-orange brick in Albany, New York. In Virginia, the English produced a pinkish-rose colored brick, as well as purplish headers for contrast, to use in their Flemish-bond masonry.

Historic Bonding and Mortar

The total replacement of mortar in the horizontal and vertical joints of a brick or stone structure is a long and expensive task. The periodic repair of portions of deteriorated mortar joints is therefore preferable. Always analyze the masonry before repointing. Many times loose bricks, cracks, or damp walls are not caused by deteriorated mortar joints, and an unnecessary and costly repair job fails to cure the condition. Mortar joints are often "washed out" by faulty gutters and downspouts causing repeated cascades of water on a specific portion of the masonry, or the dampness may be rising through the masonry walls by capillary action as result of subsurface and foundation conditions that are unrelated to mortar.

Repointing Procedures. The pattern of the masonry joints is a significant element in the architectural character of a historic structure. The traditional appearance of an old building can be radically altered by subtle changes in the width, color, and texture of the mortar joints. And even if the original width and character of the mortar joints is faithfully reproduced, there may still be a stark contrast if the old brick is not cleaned.

The new mortar should match the color of the old in a true historic restoration, and it will have to be troweled into the joints in the same style as the original. However, the composition of the mortar need not duplicate the original. Most old buildings were built with a cement-based mix that produced an inflexible mortar incapable of responding to thermal changes in the bricks. This caused bricks to spall at the edges in the summer when they expanded against an inflexible joint, and contract and pull away from the joint in winter, thus opening cracks for moisture to penetrate. Lime mortar is recommended for jointing and pointing because it compresses and flexes with thermal changes. It is also slightly soluble in water and has the ability to self-seal small cracks that might develop in the pointing process.

Since the original mortar is not being used, the color will not be the same unless extraordinary care is taken to match the lime and sand with the original materials. The easier and more practical way is to add a mortar pigment to obtain the desired color. The technique is far from new. Some late nineteenth- and early twentieth-century builders used pigments in their mortar. Choose only chemically pure mineral oxides that are alkali-proof and will not fade in sunlight. Avoid natural earths because they have poor tinting strength and avoid organically based pigments because they also fade in the sunlight.

Early nineteenth-century mortar contained lumps of incompletely burned lime and oyster shells. Modern lime is uniformly ground, so if strict fidelity to the original is required, the mason will have to duplicate the lumps of lime and shells or grind some of the old mortar and mix it into the new.

Before repointing starts, the old joints should be raked out to about one inch in depth, and further if the mortar is loose. Rake the joints with hand tools; power masonry saws inevitably slice off the edges of the bricks, causing rigid mechanical joints that are quite different from the original softly molded mortar joints. The cut at the back of the joint should be as square as possible, and all debris cleaned out with an air hose. Before repointing, the old joint should be wetted with water.

Chemical bonding agents are not recommended, and if an experimental product is to be considered, a small sample panel in an inconspicuous location should be tested first.

The inch-deep joint should be filled with mortar in layers of 1/4-inch depth, allowing time between each layer for the previous layer to reach thumbprint hardness. When the final layer is sufficiently hard, it is tooled to re-create the appearance of the original masonry joints.

Joints can be finished flush with the bricks or recessed from the face of the wall. Or they can be tooled into either a concave or a convex finish. Ideally, the original joint style should be used for the repointing, but if the edges of the bricks are damaged, a different tooling may be necessary. If the brick edges are worn, the width of the joint will be too wide and look out of scale. Also, when the joint is packed with mortar, the mortar will be spread thin at the worn edges, thereby becoming susceptible to breaking off and admitting water. A more aesthetic solution that is structurally sound, is to make a convex joint so that the mortar does not fill the space between the broken edges of the brick.

Restoring masonry is a painstaking process that cannot be rushed. An experienced contractor will be forced to bid high for such work, and it is the architect or owner's responsibility to ensure that the masons provide the craftmanship that the contract cost deserves. The original pointing may have lasted a century or more, and there is no reason why the restoration should not last just as long.

Terra Cotta

Terra cotta is an ancient material used by the Etruscans and Romans. Because terra cotta clay was abundant and easily worked, it was favored for its decorative qualities. Fired terra cotta is extremely durable, but brittle. Unglazed, its surface is porous and invariably erodes when used in building construction. Until the late nineteenth century, unglazed molded decorative terra cotta elements were generally combined with brick masonry, which means that it has to be cleaned with great care. Today, silicone coatings and other sealants help to reduce its moisture-absorbing qualities.

At the end of the nineteenth century, mass-produced glazed terra cotta facing began to be extensively used on high-rise, iron-framed structures. It served as both fireproofing and facade and was lighter and cheaper than stone because it could be molded rather than carved. Terra cotta glazes mimicked stone texture and color so successfully that many architects and owners have been surprised at the deception. Many of the earliest installations were, in a sense, experimental and have not proved to be as durable as their original builders expected. Widespread manufacturing of terra cotta ceased during the Depression, when other machine-produced materials replaced it.

Figure 7-11 Terra cotta repair, Woolworth Building, New York, New York. After more than 60 years' exposure to wind and weather, much of Cass Gilbert's Gothic-inspired Woolworth Tower was in disrepair. The Ehrenkrantz Group undertook a massive, detailed restoration program that has since restored the luster and serviceability of this venerable New York landmark.

Repairing and Replacing Terra Cotta. Restoring terra cotta is difficult. Because the liquid clay from which it is made shrinks in the firing process, plaster casts taken from undamaged elements cannot be used as molds for the clay. The mold for clay casting must be slightly larger than the desired final product. For this reason fiberglass, which is not subject to shrinkage, has become a popular and economical substitute. Since a typical terra cotta facade is made up of many different elements, each requiring a special mold, economies in the cost of casting can be significant.

Typically, nonstructural facing materials are anchored into the loadbearing walls behind them. (See Figure 7-11.) Metal ties usually extend between the horizontal joints in the wythes of the masonry wall. At the turn of the century, no rust-resistant iron or "stainless steel" was available. Weather exposure, particularly on tall buildings, has led to moisture penetration that has corroded the iron ties and has also caused cracks in the terra cotta due to freezing and expansion. Falling fragments of deteriorated terra cotta have injured and killed pedestrians in New York City. Replacement ties should be stainless steel or galvanized

Figure 7-12 Replacing terra cotta, Woolworth Building, New York, New York. In some portions of the tower the deterioration and loss of bond was so extensive that entire sections had to be removed and replaced. Because of the height and setbacks this necessitated coordination of both workers and materials.

steel. Because terra cotta cannot support many courses of material, steel shelf angles must be provided to prevent the blocks from being crushed by their own weight. These steel angles must also be inspected for corrosion and replaced if necessary. (See Figure 7-12.)

Concrete/Cast Stone

Concrete, which is a mixture of cement, sand, stone aggregate, and water, is an ancient material used since Roman times. Traditionally, it was poured into forms or used as an infilling between masonry walls. Its real advance as a building material occurred in the late nineteenth century, when iron bars were combined with concrete to form reinforced concrete. So-called historic concrete predates reinforcing and is more prone to crumbling. Modern continous foundations and footings capable of bridging uneven soil-bearing conditions require reinforcing and date from the beginning of the twentieth century. In earlier periods, massive spread footings were the only kind of foundation that could be constructed of concrete.

Cast stone (actually concrete) was the nineteenth century's answer to low-cost imitation stone facing. Many monumental public works projects were executed in cast stone and concrete. Many of those concrete landmark bridges, colonnades, park walls, fountains, staircases, and balustrades are crumbling today. Reconsolidation and restoration efforts on deteriorated cast stone are not always successful. It is often necessary to reproduce the severely damaged elements in more durable reinforced concrete or fiberglass-reinforced concrete. Now that greater recognition is being given to landmarks of engineering, such as dams, grain elevators, and mills, more attention must be paid to techniques for stabilizing and restoring historic concrete structures.

Masonry Cleaning and Coating

Cleaning techniques for one form of masonry can usually be applied to the other forms, with the notable exception of terra cotta. Overzealous restorers can do more harm than good by cleaning masonry buildings. By removing what they believe is dirt, they actually remove weathered exterior masonry "skin," exposing the softer core material and accelerating deterioration. Most cleaning efforts are undertaken for cosmetic reasons and improper procedures can lead to faster absorption of harmful environmental and industrial pollutants.

Despite its reputation for durability, masonry must be treated as a delicate material and not just a pile of stonework. Three general cleaning methods are available: (1) water, which softens dirt and rinses off deposits; (2) chemicals, which work by reacting with the dirt and/or

Figure 7-13 Facade cleaning, City Hall, Schenectady, New York. The gentlest cleaning method possible is the most desirable to prevent damage to the historic building's fabric. Combinations of marble, masonry, wood, copper, and glass require careful analysis to be certain that a successful cleaning method for one material will not be harmful to others.

masonry to hasten removal by flushing with water; and (3) mechanical abrasion followed by water rinses.

Tests should be performed, starting with the mildest, to determine the least harmful cleaning technique that provides the best results. At the same time, the technique should be evaluated for its effects on the surroundings. Will the chemicals washed off the wall harm plant life? Will airborne dust enter nearby windows? Will water seep into basements? What are the potential health effects on the workers?

Testing should cover at least one square yard and include mortar joints as well as the masonry units themselves (see Figure 7-13). Variety of material on a building may require more than one cleaning method. Some cleaning methods will be harmful to adjacent materials, such as glass or wood, and provision must be made to shield these areas.

Water Cleaning. Cleaning with low-pressure water and a bristle brush is the mildest technique. If insufficient, pressure can be increased and a nonionic detergent, such as trisodium phosphate, can be added. Both methods introduce a great amount of water into the masonry and joints. If the masonry is porous, the water can damage interior finishes, or it can freeze and thereby spall the surface of the masonry.

An alternative that minimizes water penetration uses a misting device on the hose nozzle to create a fog. The wall should be enclosed with tarpaulins to retain the fog, and fogging should be cycled to provide drying periods. One technique sprays fog for 10 seconds followed by a few minutes of drying. Other projects report cleaning marble with a nonstop eight-hour fogging.

Steam Cleaning. Steam cleaning is falling into disfavor. It is about as effective as cold-water washing and can be hazardous to the cleaning crew. However, steam is sometimes used effectively to remove sticky or greasy patches or to remove grime from decorative details.

Chemical Cleaning. Chemical cleaners have more disadvantages than advantages. Unless acids used to clean masonry are diluted to the exact recommended strength, they will etch or bleach stonework. Some acids affect polished stone or marble even in quite dilute solutions.

Companies specializing in cleaning are moving away from the common acids and are using biodegradable organic cleaners. Caustic sodium (or potassium) hydroxide is still used, but must be followed by a weak acid rinse to neutralize the masonry surface before applying the water rinse. Again, specialists temper the caustics with inhibitors to prevent them from burning the stonework. Particular stains, such as graffiti, in limited areas, can be treated with specific formulations.

Removing Paint. Painted masonry walls should be researched before stripping. Some walls were painted as part of the original design, some were painted later to improve the appearance, and some were painted in an effort to keep moisture out. (See Figure 7-14.) If the paint was added it probably should be removed, but if the paint is original, the building should be repainted. However, too many old coats of paint can cause a new coat to lose adhesion, in which case the old coatings should first be removed.

The same treatments for exterior wood can be applied to painted masonry: scraping by hand (no power tools), thermal pads, and paint solvents for architectural details.

Coating Masonry. Masonry walls seldom need coating. In fact, adding a waterproof or water repellent coating often creates trouble. Porous brick or stone seldom causes serious water penetration: therefore, if the interior of a building is damp it is probably the gutter, downspouts, or mortar joints that need attention.

A porous masonry surface admits water but it also allows moisture to

Figure 7-14 Paint restoration, Blair House, Washington, D.C. In order to choose the proper restoration method it is essential to test the existing painted surfaces in small sample patches to ascertain the nature of previous applications and the condition of the masonry beneath. Blair House was repainted after the tests.

escape through capillary action, called wicking, and evaporation. Coatings are seldom 100 percent efficient in keeping water out, and the moisture that does penetrate gets trapped and cannot evaporate because of the coating. This process occurs with moisture rising from the foundation into the wall.

Water-repellent coatings, usually transparent silicone formulations, keep liquids out but allow vapor to enter and leave. Unfortunately, the vapor can condense inside the masonry. An additional problem is that condensed vapor (water) dissolves salts from the masonry and mortar and carries them toward the surface. The salts deposited behind the transparent coating crystallize and in a few years will grow enough to push out the surface of the masonry and its coating.

In short, try to avoid coating masonry. One condition that is an exception to the don't-coat rule occurs when masonry has been sandblasted. Sandblasting can leave brickwork so porous that atmospheric pollution will cause more damage than the potential damage from the coating.

WOOD

Wood, the most traditional of American building materials, provides structural framing, siding, flooring, roofing, doors and windows, and architectural embellishments. It was widely used for a long time because trees were abundant and lumber was universally affordable. Although highly vulnerable to deterioration from moisture, properly maintained structures have lasted hundreds of years.

Causes of Wood Deterioration

Wood that is allowed to remain wet will eventually rot, as often happens when moisture becomes trapped in unventilated spaces. In some regions, termites and insects may cause considerable damage. If roofs and flashing have not been well maintained, water damage may be found in roof and cornice framing as well. (See Figure 7-15.) Decorative filigree (such as lacy Carpenter Gothic porch detailing), wood railings, fence posts, and balusters are particularly vulnerable to deterioration. On the interior, leaking plumbing or faulty ceramic tile in bathrooms can cause rotting joints and framing. Excessive humidity can cause warping in paneling and parquet floors.

During the colonial period, lead-based paints were costly and most ordinary structures were left exposed to the weather. (See Figure 7-16.) Except for wood shingles, which are still best left unpainted to allow the escape of moisture and condensation, most preservationists agree that exterior wood surfaces should be treated. Paint and wood preservatives keep moisture out of the surfaces of wood, and putty and caulk keep moisture out of joints and openings.

Figure 7-15 Wood deterioration, nineteenth-century mansard roof, Brooklyn, New York. If the normally shielded wooden roofing substructure and framing suffer prolonged direct exposure to the weather, due to the failure of the roofing material and flashing, irreversible deterioration may occur. Moisture penetration from a leaky roof will also eventually damage the framing and interior finishes.

Figure 7-16 Town house, East Hampton, New York. Ordinary buildings such as this modest eighteenth-century schoolhouse were generally left unpainted except for the milled wooden elements—windows, doors, and trim. Because of the abundance of wood, it was cheaper to periodically replace deteriorated shingles than paint them.

Badly decayed wood must be cut out and replaced, but partially damaged architectural elements and specially carved details can often be saved. The affected wood elements must first be removed, thoroughly dried, and treated with fungicide to kill any latent growths. Next, the wood is waterproofed with two coats of boiled linseed oil, with 24 hours between coats. All cracks and holes in the wood must be filled with putty, and all joints between wood members caulked to prevent water penetration. After this, the surfaces are primed and coated with a compatible paint.

Some wood preservatives such as creosote, which resists rot and infestation, also resist the application of stains and finishes. If insect infestation has progressed beyond the cellar into concealed portions of the wood frame of an old building, salvage may prove to be too costly.

Consolidation Techniques for Serious Decay

If the decay is advanced, and also in circumstances where disassembly is impractical, the affected wood can be stabilized with semirigid epoxies that saturate the wood and then harden. This new surface has to be filled with an epoxy patching compound, which, when hardened, is sanded, primed, and painted. Epoxy compounds can also be used to build up profiles of deteriorated wood instead of replacing a whole section. Conventional wood putty is also used and its long-term effects are better known than those of many of the new and still highly experimental plastic or expoxy consolidation techniques. All of these methods offer only a cosmetic treatment to the appearance of the original wood architectural elements, and not a solution for wood structural members seriously undermined by decay. When structural integrity is involved, it may be necessary to reinforce the original structures with wood or metal inserts. In some cases, replacement with framing of comparable vintage may be preferable to too-visible repairs.

Keeping Moisture Out

There are dozens of paths that moisture can take to get through an exterior wall or between the junctions of wood siding and masonry, foundations, window and door frames, and joinery such as corner boards and decorative trim. Unless the leaks are plugged, recurring exposure to moisture will undo all the restoration work.

All dried-out caulking compounds should be cleaned from the joints, which should be resealed with latex or butyl caulk. Holes in the wood must be filled with putty. Faults in the flashing can be repaired with

flashing cement. If rain gutters are leaking, they should be replaced or repaired with epoxy resin and fiberglass.

The Preservation League of New York State draws attention to the potential damage that will occur if insulation is blown into old walls without installing a vapor barrier. In cold weather, moisture may condense on the insulation and cause paint to peel from the siding. Vents in the wall may redress the problem, but restorers should try other energy-saving techniques before resorting to wall insulation.

Repairing Old Wood Windows

When a deteriorated building has significant historic value, you may want to retain the original windows. In these cases, the windows may have to be taken out of their frames and taken apart in a workshop. This enables a carpenter to remove a decayed member, such as a muntin, duplicate it with new wood, and fit the replacement into the sash. Woodworking mills in various parts of the country still make components that replicate typical sash from earlier centuries. (See Figure 7-17.) These reproductions are usually custom made, and special shapes and mullion configurations such as roundheads, fanlights, or diamond panes can be very costly. It is often worthwhile to spend some time rummaging in parts and demolition warehouses before commissioning custom millwork. If it is impractical to make exact replicas, it may be better to salvage the restored old sash on one facade and group the new reproductions on another, less prominent elevation.

Figure 7-17 Eighteenth-century window, Connecticut. Because of the austere simplicity of Early American buildings, the windows are often the most significant exterior feature. Their spartan elegance—often derived from minor elements such as subtlety of proportion and detailing of muntin profiles—can be lost through clumsy modern window replacement.

Windows are particularly susceptible to the ravages of rain and snow. Because the component parts of a window and its frame are relatively slender, they should be treated with a water-repellent preservative or a simple water-repellent solution. The U.S. Department of Agriculture's Forest Products Laboratory compared both methods in a 20-year exposure text and concluded that they are of equal merit. Windows without treatment began to show deterioration after only six years. The laboratory's recommended water-repellent formula calls for mixing three cups of exterior varnish with one ounce of paraffin wax and adding mineral spirits, paint thinner, or turpentine to make one gallon. This solution can also be liberally brushed on woodwork, allowed to dry for three days, and painted. Tests show that paint lasts longer on treated surfaces than on untreated surfaces. Old paint should be removed, using the techniques described later in this chapter under "Paint."

Improving the Energy Efficiency of Old Windows. Because of seasonal variations in humidity, all woodwork is subject to dimensional variations—shrinking in the winter and swelling in the summer. Current energy considerations require weather stripping to be installed around windows to prevent cold air infiltration. Felt strips have been traditionally used, but because they hold moisture they may contribute to the deterioration of the wood sash. That is why vinyl is now preferred. The benefits of adding sash locks to the meeting rails of double-hung sash to improve thermal efficiency is another minor and worthwhile concession to historic authenticity.

The same logic holds true for storm windows. To retain the original appearance of the building, unfinished alumnium storm frames should be avoided. Frame colors of metal storm windows should match the building's trim color. The Technical Preservation Services of the National Park Service recommends not putting storm windows inside the old windows because moisture trapped between them condenses on the inner surface of the exterior glazing and leads to deterioration. The best solution is to place unobtrusive modern storm windows outside the original windows. If glazed with unbreakable glass or Lexan, the new windows provide a security barrier for old glass. The Technical Preservation Services advocates retaining and repairing original windows instead of replacing them with copies. And, it adds, a storm window outside a reconditioned original window provides better thermal insulation than a new double-glazed metal window.

Figure 7-18 Voorlezer's House, Richmondtown Restoration, Staten Island, New York. A restored eighteenth-century rural Dutch schoolmaster's house has rough-and-ready plank shutters. More elegant town dwellings of the same period usually have more refined paneled or louvered exterior shutters. During the colonial period shutters were a functional element for weather protection and security.

Figure 7-19 Guernsey Hall, Princeton, New Jersey. A mid-nineteenth-century Italianate Villa retains its original painted terne roof in a sensitive adaptation to apartments by architects Short and Ford.

Wood Shutters

During the colonial period most domestic buildings had wood shutters, often with a duplicate set inside and out. (See Figure 7-18.) This tradition continued through the eighteenth and nineteenth centuries and has been perpetuated in the twentieth century on Colonial Revival style dwellings. Because of their constant exposure to the weather, shutters are often the first item of wood trim to deteriorate. Generally, old shutters on the lower story were solid panels and the upper stories had "blinds" with adjustable louvers. It is common to observe louver slats dangling or missing from the blinds. Evidence of long-missing shutters can usually be found by looking for the remaining iron hinge plates and hold-back devices or traces of hardware mounting holes. Replacing missing shutters is not easy, since few survive in good enough condition to reach parts warehouses. Replacement shutters stocked by most lumberyards are rarely proportioned correctly for old buildings so that period reproductions have to be custom made. Pressure-treated wood offers greater weather resistance. Replacement of missing shutters, as well as traditional awnings, can contribute significantly to the energy efficiency of an old building in cold weather. Louvered doors backed with metal screening are an attractive alternative to modern screen entrance doors.

METAL

Metals have long been used in sheet form for roofing, in cast form for facade elements and decorative details, and for structural support. (See Figure 7-19.)

Corrosion

The drawback to metals is that they oxidize, that is, revert to their natural ores. For instance, iron turns into iron oxide and in doing so undergoes a change in the physical condition of its surface. In two

Figure 7-20 Queen Anne row houses, New York, New York. In the late nineteenth-century painted pressed metal decorative parapets, bay windows, dormers, cornices, and shopfronts were often used as an inexpensive alternate for repoussé copper.

words, "iron rusts." Tin, which had been used in sheet form on roofs, was one of the first materials to be dipped in other metals, such as lead and zinc, to provide a protective coating against atmospheric corrosion.

The atmosphere was much cleaner in the nineteenth century and metals on buildings were less harmed than in this century. Now, unfortunately, motor vehicle exhaust products and other airborne toxic pollutants are deteriorating historic buildings at an accelerated rate. Century-old metal is often ravaged beyond repair. Corrosion from atmospheric pollutants is the most ubiquitous cause of deterioration. Sulfur compounds liberated when fossil fuels are burned are absorbed in atmospheric moisture, enveloping metal on buildings with an acid moisture, called acid rain, that starts an electrochemical action.

In a similar manner, buildings along seaboards suffer from the corrosive action of chlorides and salt particles from the marine atmosphere. Even serviceable modern materials, such as aluminum, develop surface pitting near the ocean. Chlorides have the same effect as sulfides from smokestacks.

The electrochemical action that eats away the metal's surface can also be caused by two different metals in contact in the presence of moisture. When this occurs, the metal with the lower electrical potential deteriorates the most. All metals and alloys are ranked by degree of potential. From the rankings, one can ascertain which of the two metals will deteriorate first, and from the size of the difference in potential, one can estimate the intensity of corrosion. One of the most commonly seen examples of this form of electrochemical corrosion is caused by mistakenly patching an old copper roof with aluminum.

Corrosion occurs in several types of metal connectors used to hold metal or masonry ornaments and details to a building's structural system. When these connectors corrode they may fail with disastrous results. If even one corroded connector is found, it may be necessary to replace all the connectors in the area with a more corrosion-resistant material.

Figure 7-21 Bay window, Brooklyn, New York. Elaborate, projecting architectural elements such as bay windows and cornices were traditionally constructed of wood to support the light-gauge pressed painted metal. If the exterior painted metal is accidentally dented or pierced, water may penetrate the hollow interior spaces and cause rust and deterioration.

Stress Cracks

Stress cracks in metal are vulnerable to corrosion. Metals that have been worked into shape by pulling, bending, or machining contain minute stress cracks that caustic or ammonia solutions will enter and enlarge. Sometimes this action causes pieces of architectural detail to break off. Light-gauge pressed metal, usually tin or copper, used for bay windows, decorative dormers, and cornices, becomes brittle with age and is easily punctured. (See Figure 7-20.) Careless placement of a painter's ladder or scaffold can cause serious damage. Once water penetrates behind the pressed metal, corrosion begins to undermine the iron or wood supporting armature and may eventually cause the collapse of the architectural feature. (See Figure 7-21.)

Expansion/Contraction Factors

Metal failure occurs when metal cannot freely contract and expand with temperature changes. Metal roofing tends to creep under the combined effects of high temperatures and its own weight, especially with heavy material such as lead. Creep causes the metal to wear thin in some areas, and then it ruptures. When an affected area is renewed, expansion joints should be built in, to apportion the roofing into smaller sections.

Abrasion

Metals also erode by abrasion. When rain washes fragments of deteriorated slate roofing down roof valleys and gutters over time, metals wear thin. With interior metals, erosion is most often seen on railings, push plates, brass thresholds, and similar fittings that have constant traffic.

Figure 7-22 Portico, Savannah, Georgia. A handsome nineteenth-century small decorative roofed balcony has been fabricated of cast-iron elements. Although cast iron was sometimes combined with wooden structural framing, this is not apparent once painted.

Other Mechanical Repairs

Worn areas on sheet metals can be patched with "liquid metal" if the hole is small, or larger areas can be patched with a piece of the same material. The patch can be applied with heat, such as in soldering, or it can be connected with fasteners like rivets.

Architectural details that have been cast or wrought can be replaced if a piece is removed for a mold to be taken and a new metal copy cast. If the damaged detail is small enough, it can be built up with liquid metal.

When structural members have deteriorated, the restorer should stand back and leave the repair work in the hands of competent engineers and steel fabricators. Temporary supports will be needed for floors and roof, and, in some cases, exterior walls will have to be braced, while the members are spliced or reinforced.

Conserving Metal

Many preservationists believe that repairing metal on a building is too expensive, and that the less costly alternative is to remove failed metal and replace it with a new material. The replacement can be the same type of metal that has been specially treated to delay corrosion, or it might be a new material, such as polyester or reinforced glass fibers.

Although metal can be protected in several ways in a factory, not all of these techniques are applicable to historic metal elements in the field. Whereas in the factory sheets can be coated with ceramics or with metals such as zinc, tin, and lead, in the field they can only be coated with organic paint. Metal roofs should be painted and never covered with conventional roofing coatings made from asphalt or bitumen because the interaction will only accelerate deterioration of the metal.

Cast Iron

In the late eighteenth century, iron elements such as columns and arches were incorporated in masonry construction. (See Figure 7-22.) By the latter half of the nineteenth century whole cast-iron fronts were applied as facades to commercial buildings that were otherwise of masonry construction. Some corner buildings had two elaborate cast-iron facades, and occasionally, freestanding cast-iron framed structures were designed as pavilions or greenhouses. Cast in sections, these were specimens of some of the first prefabricated building systems.

Architectural cast iron usually comprises small sections bolted together, with joints and connectors caulked to prevent water penetration. If the caulking does not dry out and if the surfaces are kept painted, the metal should last forever. Unfortunately, there are many physical forces, such as subsiding footings or poor design details, that trap water and cause rust to form, resulting in deterioration of the metal.

Repairing Damaged Cast Iron. Iron can be patched with plumbing epoxy or auto putty, or it can be welded with nickel-alloy welding rods. Welds should be ground smooth and not left as standing seams. All repairs should be preceded by an attempt to rectify the cause of the problem. Railings and steps frequently fail because water penetration causes the bolts to rust. New bolts must be installed and tightly caulked before the cast-iron components can be restored.

Stamped metal panels used in the upper stories of buildings are much thinner than the street-level cast-iron facades and, therefore, they cannot be treated as roughly. Many of these thin panels are worn, and the holes must be patched to prevent water from getting behind the panels. Small patches can be made with "liquid metal" used by auto mechanics. All seams should be checked to prevent leakage. Then the metal should be primed and painted as described below.

Cleaning and Painting Cast Iron. When cast iron is renovated, it should be cleaned of rust, grime, and loose paint. Cast iron is about the only material that can be sandblasted. It can also be cleaned with rotary wire brushes, scrapers, heat pads, chemical removers, and propane gas guns. It must be stripped to a firm layer of paint. Fluted columns or fragile details on column capitals may require handscraping and removal of all previous coatings so that the delicate designs will be evident. Balconies and railings will look better if they are first cleaned down to the bare metal. Chemical removers work well for this task.

The Friends of Cast-Iron Architecture recommends that two coats of rust-retarding primer, such as red lead or zinc chromate, be applied immediately after the metal has been degreased with alcohol. Joints and bolt holes in cast-iron facades and ornamental ironwork should be caulked after the first primer coat. Primers should be followed by a generous coat of enamel paint that is brush, not spray, painted. The enamel seals and protects the primers. (See Figure 7-23.)

Figure 7-23 Cast-iron columns, Grey Art Gallery, New York, New York. When commissioned by New York University to transform a street-level study hall facing Greenwich Village's Washington Square into an art gallery, architects William C. Shopsin and Giorgio Cavaglieri chose to emphasize the massive turn-of-the-century Doric cast-iron structural columns by painting them white to contrast with the neutral background of the carpet, ceiling, and display-wall surfaces.

Galvanized Steel

Zinc is seldom used in its pure form, but it is the basic protective element in galvanized steel sheets. Zinc is vulnerable to corrosion from the atmosphere, cement, wood, plants, and condensation. Its virtue is that when penetrated, the steel beneath it remains protected because the zinc serves as a sacrificial metal.

Galvanized metal is difficult to repair. It must be soldered with care because acid flux can dissolve zinc. It can be painted after weathering for several months. Rusted areas can be treated. Many project directors prefer to replace galvanized sheets rather than repair them.

Architectural details, such as Corinthian capitals assembled from stamped or pressed sheets of galvanized steel usually have to be disassembled if they have deteriorated. Missing pieces can be replaced and soldered into position. Then after cleaning, the components can be put through the galvanizing process and recoated with zinc. Fiberglass reproductions can be made of painted pressed-metal ornaments or of the verdigris patina of old repoussé copper.

Copper

Copper is the major ingredient of brass and bronze and is used in pure form as a durable roofing material. Copper sheeting acquires a green patina that serves as a protective coating. But copper is vulnerable to failure if it is not allowed to expand and contract, is subjected to abrasion, or has insufficient support (Figure 7-24).

Figure 7-24 Repoussé copper oval wreathed dormers, mansard roof, New York, New York. At the turn of the century, roofs were complex confections of slate and repoussé copper flashing. Weather exposure and atmospheric pollution have hastened normally slow deterioration of copper, causing it to become brittle.

If copper sheets have to be replaced, the new material should not exceed 8 feet in length, should be laid on rosin paper to facilitate thermal movement, and should be fastened with copper clips and nails. If the original sheathing boards are spaced apart they should be replaced with tongue-and-groove boarding that supports the copper.

Sheets weighing less than 20 ounces can be soldered with 50 percent pig lead and 50 percent tin solder. Sheets weighing more than 20 ounces require copper rivets for all connections.

Occasionally the surface copper erodes, and this can be redressed by painting with red-lead primer and white-lead finish coats. Alkyd resin paints formulated for copper are also available. Before it is painted, the copper must be cleaned with a solution of copper sulfate and nitric acid to remove grease and oil from the pores.

Copper Alloys

Brass and bronze should be cleaned only when absolutely necessary, since their patinas protect the metal. Salts and bird droppings can be removed with water and a soft brush.

When repaired pieces are riveted or brazed onto the original metal, the patina has to be removed with one of several chemical formulations. The Copper Development Association recommends a 5 percent oxalic acid and water mixture and finely ground pumice rubbed in with a soft cloth and rinsed well. Neither bronze nor brass should be sandblasted, but sculptures have been blasted clean with minute glass beads or crushed walnut shells. These processes require an expert hand and should not be attempted by a general restoration contractor. (See Figure 7-25.)

GLASS

In the early colonial period, glass was an imported commodity affordable only by the privileged. Early window glass was hand-blown and then cut into lights of limited size. Some seventeenth-century New England settlements continued the medieval tradition of leaded diamond pane casements. By the eighteenth century, 6-over-12 and 9-over-9 lights were common proportions. The knob that remained where the glassblower attached his pipe, called a bull's-eye, was sold for transoms over entrance hall doorways because it was less desirable than the flat sheets but too precious to be thrown away. (Nowadays these bull's-eyes are coveted and imitated.)

As domestic glass production increased and technology improved, the

Figure 7-25 Beaux-Arts windows, Dorillton Apartments, New York, New York. At the turn of the century, the fascination with picturesque architectural features led to many eclectic window designs. If elements of these curved or intricate millwork novelties are deteriorated or missing they are costly to reproduce.

size of glass lights also increased. By the beginning of the nineteenth century 6-over-6 was the most common pattern. By mid-century, 2-over-2 was favored, and finally by 1900, 1-over-1 was popular. Because window sashes were often changed, the pattern of lights is not an absolutely reliable guide for dating old buildings. Various historic revival and eclectic styles, such as Queen Anne, have odd windows that do not conform to these prototypes. As a general rule, however, the earlier the building, the fewer and smaller the windows in comparison to the amount of wall area.

By the end of the nineteenth century fairly large sheets of glass were available for shop windows, but only in the twentieth century were huge sheets of plate glass, curtain walls, and structural glass available.

Figure 7-26 Nineteenth-century wood storefront, Newton-Sussex, New Jersey. The glazing elements of most typical nineteenth-century storefronts were fabricated of wood and then painted, including those set in cast-iron facades. Glass sizes for these display windows were limited by the technology of the period.

Glass is mainly used for windows, skylights, and conservatories, but it has also been used to form structural blocks and facade veneers. It is impermeable and resists ordinary atmospheric corrosion reasonably well. But it has no resistance to impact, to scratching, or to some chemicals. Historic glass, made before the technology was improved, often contains chemicals that react under prolonged ultraviolet light to produce discoloration. Old glass contains visual imperfections and air bubbles, and it may be bowed and brittle. Its brittleness makes it very difficult to cut so that it may not be possible to cut smaller lights out of older ones. If old lights are mixed with new glass, the difference will be very obvious. It may be necessary to combine old lights of the same size from separate sash to achieve a unified window. Some stained glass dealers stock handmade reproduction glass that imitates the imperfections of old glass. Although costly, it is not as brittle and is more serviceable.

Protection

Glass of any age must be protected when a building facade is cleaned. Acids and chemical paint cleaners used on masonry will etch or scratch glass. Also, glass panes can be easily broken when an attempt is made to open a sash that has been painted shut for years.

An old building is most vulnerable in the midst of construction. Because it looks unoccupied it becomes a natural target for vandalism. If there is any surviving historic or special glazing it should be protected against breakage; if at all possible, it should be removed for safe storage during the construction process. Even if windows are adequately protected from the exterior, damage can result from carelessness during interior demolition. Old glass that is loosely set or in severely deteriorated window frames is not likely to remain intact (Figure 7-26).

Restoring Glazing

Putty, a mixture of linseed oil and chalk, keeps water and air from entering a window around the edges of individual lights. When glazing putty dries out, water seeps between the glass and the sash. In that case, remove the old putty and install new flexible putty. To make a secure seal, the glass should be seated in putty so that the putty is behind the glass and separates it from the sash. Make sure there is a 1/16-inch space around the panes to allow for thermal changes or movement of the building. All putty should be protected with paint to seal out water.

Removing old putty can be difficult if it doesn't break away with scraping. Putty that adheres to the wood has to be removed with chemicals or heat. Both methods have drawbacks. Alkalis do break down the linseed-oil binders in putty, but the disadvantage is that they also raise the grain of the wood. Alkalis must be thoroughly washed away and the wood neutralized with weak acid; otherwise, the new putty won't adhere. Paint removers or lacquer thinners will soften the residual putty enough for scraping. The alternative is to use an electric heating wire on the putty. To minimize the risk of cracking the glass with heat, both sides of the glass should be heated with electric light bulbs and, when hot, the side being worked on should be protected with asbestos before using the hot wire. This work is best conducted horizontally rather than in a window frame. Torches and heat cannot be used to remove paint on wooden sash containing decorative lead or soldered glazing channels, as in some fanlights.

Cleaning Glass

If old glass has not been regularly cleaned, an oily film from burning fossil fuels builds up on the glass. The film can only be cleaned with very fine 0000 steel wood and detergent. Do not use a coarser grade of steel wool, and do not press too hard on the glass, or permanent scratches will result.

Figure 7-27 Nineteenth-century greenhouse, Brooklyn, New York. The technology used in early greenhouse construction is now obsolete. Thus attempting to replace old iron framing with wood glazing members is not very practical. Reproductions fabricated of cast aluminum and vinyl channel glazing systems are far more serviceable.

Curved Glass

The eclectic architecture of the late nineteenth century abounded in turrets, orioles, and bay windows whose sensuous contours required curved windows with curved panes of glass. To form curved glass, a sheet of flat glass is laid on a steel form bent to the required curvature and heated in a kiln. The glass becomes plastic and drapes to the profile of the form. The steel forms are expensive, since most are custom made. Some glazier's shops have standard forms, and they can supply less expensive service if your project's requirements conform to stock sizes. Many buildings with broken curved glass have been restored with curved sheets of clear Plexiglas because custom-bent glass is brittle and expensive.

Owners seeking greater energy efficiency often want storm windows or double glazing installed on their drafty old houses. Most of the stock extruded-aluminum profiles used for storm windows cannot be bent to fit the radius of old curved windows so the only alternative is custom-made Plexiglas or Lexan. Condensation, fogging, and cleaning are problems in these situations.

Greenhouses

Greenhouses were made of solid cast-iron or tubular pipe with inset wood glazing stops until after World War II (Figure 7-27). Because of the constant high humidity and the resultant condensation, most of these old framing systems are rusted and deteriorated beyond repair. Curved or specially formed glass was custom bent to fit the domes and cupolas of the great nineteenth-century conservatories.

Similar difficulties are also encountered in restoring the protective glass roofs that were erected over large, decorative interior ceilings of stained glass. New insulated-and-shatterproof glass is now manufactured for modern greenhouses. Reproductions of original cast-iron profiles can now be cast in nonrusting aluminum.

Glass Block

Solid glass "plugs" were being produced by the mid-nineteenth century for insertion in cast-iron library-stack floors, shop stairways, and sidewalk vault areaway covers. Glass "bricks" or hollow blocks became popular after World War I. Modern architects of both the Art Deco and International styles became quite enamored with the possibilities of glass block. Contemporary architects have become so intrigued with glass block that reproductions of the 1930s and 1940s types are once more available. (See Figure 7-28.)

Figure 7-28 Lescaze town house, New York, New York. During the 1920s and 1930s architects became fascinated with glass block as a building material. A typical nineteenth-century brownstone row house was modernized by a Swiss architect in an international style, with large areas of glass block providing daylight and privacy.

Checklist 7-1
POTENTIAL PROBLEMS

STRUCTURAL AND FACING MATERIALS

Mortarless Fieldstone or Rubble Wall Foundations
- ☐ Needs adequate site-drainage systems to avoid leaky cellars.
- ☐ May require stabilization by underpinning.
- ☐ May require excavation to increase headroom in cellar.

Irregular Mortar Joints in Rubble Walls
- ☐ May be subject to moisture penetration.
- ☐ May be difficult to trace source of leaks if whitewashed.

Brick with Mortar Joints
- ☐ May require replacement of cracked or broken bricks.
- ☐ May require repointing of joints in same style and coloration as original pointing.

Brownstone or other Cut-Stone Facing
- ☐ May be seriously deteriorated by atmospheric pollution.
- ☐ May be discolored by rusted metal elements.
- ☐ May be difficult to match original material.

Cast Stone
- ☐ May require replacement of crumbling elements, using reinforced concrete.

Unglazed Terra Cotta
- ☐ May require sealant to reduce moisture-absorbing qualities.
- ☐ May require replacement of eroded portions.

Glazed Terra Cotta
- ☐ May require replacement of cracked brittle portions, using molded fiberglass or polymer concrete.
- ☐ May require replacement of deteriorated metal ties and/or shelf angles.

Stained or Soiled Masonry
- ☐ May require cold-water or fog cleaning; test area, including mortar joints, to evaluate effects.
- ☐ May require careful use of steam cleaning to remove sticky or greasy patches.
- ☐ May require specific chemical formulations to remove particular stains or graffiti in limited areas.

Painted Masonry
- ☐ May require removal of old paint if not part of original design.
- ☐ May require removal of extraneous coats of paint and/or repainting consistent with original design.

Sandblasted Masonry
- ☐ May require water-repellent coating to prevent further damage from atmospheric pollution.

Wood Structures and Facings
- ☐ May require treatment with water-repellent preservatives or paint to prevent moisture penetration of surfaces.
- ☐ May require putty and caulk to keep joints and openings moisture-free.
- ☐ May require fungicide treatment to kill latent growths on damaged elements.
- ☐ May require removal and replacement of badly decayed elements.
- ☐ May require reinforcement of structural elements.
- ☐ May require stabilization of decayed elements, using putty or semirigid epoxies.

Metal
- ☐ Deteriorated structural members may require additional reinforcement.
- ☐ Corroded connectors should be replaced by corrosive-resistant ones.
- ☐ Light-gauge metal may contain stress cracks, causing decorative details to break off.
- ☐ Small areas may be patched with liquid metal, solder, etc.
- ☐ Cast or wrought details can be replaced with molded copies.
- ☐ Cast iron can be patched with plumbing epoxy or by welding; replace rusted bolts; clean, prime, and repaint if necessary.
- ☐ Galvanized steel may require disassembly for repairs because acid flux can dissolve zinc and make recoating necessary.
- ☐ Brass and bronze require chemical removal of patina prior to riveting or brazing.

Glass
- ☐ If adhesive has deteriorated, remove and recement panels of structural glass verneer.
- ☐ Remove stained glass panels to repair deteriorated leading or to replace broken pieces of glass.
- ☐ Use handmade reproduction glass to replace brittle historic glass.
- ☐ Use Plexiglas to replace broken curved glass.
- ☐ Replace dried-out putty in windows with new, flexible putty.
- ☐ Clean oily film from window glass with very fine steel wool.
- ☐ Replace deteriorated cast-iron greenhouse frames with nonrusting aluminum reproductions.

REPAIRS TO ROOFING MATERIALS

Wood Shingles
- ☐ Remove damaged shingles and slide in replacement shingles (suitable for small areas of damage).
- ☐ Coat underside of loose or new shingles with roofing cement.
- ☐ Use galvanized nails and cover heads with roofing cement.
- ☐ Consider replacing original roofing with new pressure-treated, fire-resistant shingles.

Slate

- ☐ Repair cracked slate by working roofing cement into cracks.
- ☐ Install new slate with copper or zinc-coated nails.
- ☐ Use slate of matching thickness and coloration.

Clay Tiles

- ☐ Glazed terra cotta may require custom-made replacements for missing or damaged pieces.

Asphalt Shingles

- ☐ Match weight of replacement shingles to that of original ones.
- ☐ If entire roof is to be replaced, consider using heavier weight to prolong life.
- ☐ If shingles have been installed over damaged wood shingles, remove both roofing materials.
- ☐ Recement curled edges with roofing cement.
- ☐ Recement torn shingles and nail down along both sides of tear.
- ☐ For a wide tear, insert a strip of copper beneath the shingle before cementing and nailing.

Built-Up Asphalt

- ☐ Fill small cracks with roofing cement.
- ☐ For larger cracks, cut out damaged area; insert and nail patch of roofing felt; cover with roofing cement, another felt patch, and fine gravel.
- ☐ Cut blisters open and repair as a large crack.

Copper Roofing

- ☐ If copper is eroded, clean chemically and paint with red-lead primer and white-lead finish.
- ☐ Replace spaced sheathing with tongue-and-groove boards to provide adequate support.
- ☐ Replacement sheets should not exceed 8 feet in length and should be laid on rosin paper to facilitate thermal movement.

Lead Roofing

- ☐ Can be coated with copper or painted after surface cleaning with Versene powder or acid.
- ☐ Patch small holes by inserting lead and melting edges.
- ☐ For widespread deterioration, consider replacement with hard alloy of lead.
- ☐ Replacement sheets should be laid on rosin paper and should have expansion joints.

Terneplate

- ☐ Paint original terneplate with red-lead or iron-oxide primer and two coats of oil-base, high-gloss paint.
- ☐ Clean and resolder opened seams and popped-up nailheads; refasten sheets with galvanized iron or steel nails or tin cleats.
- ☐ For extensive replacement, consider using terne-coated stainless steel or lead-coated copper, which do not require painting.

Flashing, Gutters, and Downspouts

- ☐ Patch small holes in flashing with solder or roofing cement; patch large holes with metal patches of original material.
- ☐ Renail or recement loose flashing.
- ☐ Clean, prime, and repaint gutters and downspouts.

PAINT, PLASTER, AND PIPING

Paint on Wooden Surfaces

- ☐ Remove dirt, grime, soot, and chalking by washing with household detergent.
- ☐ Remove mildew with a bleach and nonammoniated detergent; use mildew-resistant primer and finish coats to repaint such areas.
- ☐ Prevent rust stains by hand sanding metal parts, countersinking nails, and filling nail holes with wood filler.
- ☐ Scrape and sand crazed, peeling, blistered, or wrinkled areas requiring limited paint removal.
- ☐ To correct peeling or cracking caused by moisture, strip paint to bare wood using thermal or chemical methods.

Plaster

- ☐ Apply chemical treatment to mud-plaster walls to prevent fungal infestation.
- ☐ Check for cracks in stucco walls subjected to freeze-thaw cycles.
- ☐ For small areas of bulging plaster, chip out the plaster and fill hole with patching compound or new plaster.
- ☐ For larger areas, drill through the plaster, insert screw, and refasten the lath to the stud; if lath is missing, remove damaged section of plaster and build up new plaster on wire mesh.
- ☐ Fill cracks in plaster, cover with fiberglass tape, and coat with flexible patching compound.
- ☐ Cover crazed areas with fine scrimcloth before painting.
- ☐ Cut out damaged sections of ceiling decorations and replace with newly cast reproductions.

Piping

- ☐ Replace lead or cast-iron pipes with copper ones.
- ☐ Replace galvanized steel pipes if deposit build-up equals half the pipe's diameter.
- ☐ Make sure no plastic pipes were illegally installed.
- ☐ Make sure galvanized pipe is not directly connected to copper pipe.
- ☐ Make sure vents and traps on waste lines meet current codes.
- ☐ Make sure gas pipes for sconces have been properly disconnected.
- ☐ Make sure pipes are large enough for new fixtures and for fire-protection system.

Glass Facing

During the 1930s and 1940s, structural glass (a misnomer since it is not load-bearing) was in vogue as a veneer for storefronts and bathrooms. The opaque glass was made in large sheets, in thicknesses from 3/8 inch to 1 1/4 inch, and in a variety of colors. Although the glass resisted wear, staining, and dimensional changes, it was not resistant to impact and was subject to breakage. This is one reason why many facades were later torn down and why relatively few survive intact. Today it is difficult to obtain the original range of colors for repairs and replacement, although black-and-white Carrara glass is still available.

Structural glass is applied as a veneer over a solid wall, and was popular as a method of "modernizing" older brick-front stores. The glass is applied to the solid wall with daubs of an asphalt-based mastic adhesive, which can deteriorate with age. If a glass veneer facade is to be renovated, the condition of the adhesive should be checked, and if necessary the panels should be removed and recemented into place.

The masonry or stucco wall surface and the back of the glass must be cleaned of old adhesive and grime that has infiltrated. The bottom edges of the panels sit on at least two shelf angles, and should be inspected for corrosion. Replace them, if necessary, but note that the glass should project forward of the angle so the metal can be covered with caulking. Two brackets are needed for any panel over 15 inches wide, and they should be spaced no more than 18 inches apart. Unless the panels are small, brackets should be installed for every course of panels.

The asphalt-based mastic adhesive does most of the work in supporting the glass veneer; the daubs should be applied in a molten state. A wide band of mastic must be applied around the perimeter of the panel. This band and the daubs should equal half the panel's area.

When pressed against the mastic, the panel will have enough movement for it to be plumbed and leveled. Apart from sitting on the brackets, the glass should not be in contact with any adjacent panels or other building elements. Joints between panels are about 1/32 inch on vertical edges and 1/16 inch on horizontal edges. Keep glass 1/8 inch to 1/4 inch apart from other materials. All joints should be filled with flexible caulking compound.

Glass panels in toilet stalls are used in a similar manner to marble panels. The construction details are straightforward, and if a toilet stall needs renovating, the restorer will be able to dismantle the partitions and re-erect them in the original sequence. Some of the panels are heavy, since 1 1/4-inch thick glass was used from floor to ceiling. Other panels are composed of two 7/16-inch sheets cemented together with steel reinforcing straps between them.

The most difficult part of restoring structural glass is finding replacement panels. The manufacturers stopped production about 25 years ago, so the restorer has to search junkyards. Some plastics make a reasonable match, but they are best used when the glass panels are completely removed and the plastic can be a substitutive material.

Figure 7-29 Stained glass, Ballantine House, Newark Museum, New Jersey. Decorative stained glass or "art glass" with allegorical motifs reached the height of its popularity in the late nineteenth century. Even though the glass itself may endure for centuries, years of weather exposure weakens much of the original leading and frames, requiring skilled restoration and repair.

Stained Glass

Because stained-glass and leaded-glass windows are composed of many small pieces, they are frequently in need of repair. (Figure 7-29.) As a result of the tremendous revival of interest in crafts there are many books available on fabricating and repairing stained-glass windows. Fabrication and repair are tasks best left to a specialist. Maintenance, however, can be accomplished by a careful person. The glass can be washed in place with a solution of ammonia (no more than 5 percent) and a soft bristle brush and then rinsed thoroughly and dried with soft cloths. If the leaded glass has painted pieces you may have to use water only. Painted sections are always on the inside and, therefore, are less dirty. The putty between the leaded work and the wood frame can be replaced, and if the lead between the pieces of glass is oxidized it can be scraped clean and then coated with linseed oil.

Missing and cracked pieces of glass, or a bowed section, can only be worked on after the entire panel is removed. Before doing this, document the material with color photographs. Then entrust the work to an experienced craftsman. Once removed from its original frame, the window can be disassembled to repair deteriorated leading or broken pieces of glass. When the entire stained-glass panel has been repaired it can be fitted with an exterior sheet of a shatterproof material such as Lexan. This exterior sheet not only minimizes the risk of breakage but also acts as insulator.

Figure 7-30 Clay tile roof, Sever Hall, Harvard College, Massachusetts. Restoration of one of H.H. Richardson's most celebrated landmarks is a complex process, requiring careful disassembly of architectural features. This tympanum and cornice involved the intersection of several materials: roofing tiles, metal flashing, and brick.

ROOFING

Roofing is more likely to need replacing than any other component of a building. And unless the roof is secure from moisture penetration, it is foolish to begin any other serious restoration.

Evidence of roofing failure usually shows up in the interior before it is visible on the exterior. Damp patches, flaking paint, and actual dripping water on walls or ceilings are the signs of roof trouble, but they are not clear signs. You will have to determine whether it is the roofing material itself, the flashing, or the gutter system that has failed. Conventional practice says that if more than 10 percent of a roof area has deteriorated enough to need replacement, then the whole roof should be replaced.

With a historic building, all reasonable efforts should be made to preserve the original fabric, which means not being hasty to replace the whole roof. One solution, patching, may be practical if a similar material can be found. Sometimes it is possible to salvage enough original material from less prominent areas to restore the roof of the principal facade or from dormer roofs which, because they are seen in different planes, do not broadcast slight variations in color or texture of new material.

Do not replace any roofing until the roof structure beneath it is certified as sound (Figure 7-30). Visually inspect for a sagging ridge, drooping or rotted rafters, and deteriorated sheathing. Then take measurements to check the building's alignments and probe for soundness in the structural timber. If all is found to be well, closely examine the material in the roofing. One indication of trouble is roofing material that remains wet long after a heavy rainfall.

Types of Roofing

The earliest non-native roofing materials in America were small units, called shingles or tiles, made of wood, clay, or slate laid on pitched roofs. Early roofing materials made in large units, such as rolls of lead, copper, or terne-coated steel, were applied on shallow pitched roofs, domes, and dormers. After the turn of the century, pressed metal tiles, asphalt tiles, and asphalt-impregnated felt rolls were used on pitched roofs and on flat decks.

Wood Shingles

Wood shingles, the most popular indigenous roofing material, is far less impervious to the effects of weather than other roofing materials (Figure 7-31). Also, wood shingles require regular maintenance, particularly to remove moss, fungi, and leaves that retain water. Trees that shade a wood shingle roof should be cut back so that the shingles can dry out after a rainfall. Wood shingles generally last about 30 years, although some roofs have lasted a century. Early wood shingle roofs resemble modern ones in having a typical weather exposure of 8 to 12 inches. Old tapered shingles were as long as 36 inches, so that only about one-third of the shingle was exposed and the combined thickness at any given point was considerably more than contemporary installations. Thus, the distinctive original character of exposed gables will be lost if new shingles are shorter. The widths of the early wood shingles also varied, creating a handsome texture and shadowline, and relieving the often-severe simplicity and plainness of the pure geometric building forms.

Wood shingles can be replaced in small areas, but it is difficult be-

Figure 7-31 Dutch Colonial house roof restoration. A well-intentioned restoration imposed a heavy slate roof on an old house, overburdening the original framing. The slate has been replaced with wood shingles which are lighter and just as historically appropriate.

cause the shingles are covered by other shingles in the next course above. Damaged shingles should be split and removed in pieces. Then the nail under the shingle one course higher must be cut so that a new shingle can slide under it. The bottom of the replacement shingle must be coated with roofing cement before it is installed, and after it is nailed in place the hot-dipped galvanized nailhead must be covered with the same cement. Loose shingles can be renailed after first coating the underside with roofing cement. Also, cover the new nailhead with cement before hammering.

One compelling reason to replace original roofing material is to improve its fire-safety rating. There are now wood shingles that are pressure-treated with fire-resistant materials. The product is more costly than standard wood shingles, but its use significantly reduces fire hazards in an all-wood structure.

Figure 7-32 Slate roofs, Providence, Rhode Island. The nineteenth-century delight of pattern and color is well illustrated in this turreted and terraced hilltop house with its elaborately patterned multicolored slate roofs and iron decorative cresting and finials. Restoration of slate roofs of this type is feasible but can be very costly.

Slate

Slate quarries began operating in the United States at the end of the eighteenth century, and the material became popular by the mid-nineteenth century. (See Figure 7-32.) It is available in several shapes—round, hexagon, octagon, diamond, and rectangular—and a variety of colors. These shapes enable roofers to create interesting patterns on pitched roofs. During the 1920s and 1930s, thick rough-cut, irregularly shaped slate was in demand for revival styles such as Tudor, French Norman, and Colonial.

Quarries still manufacture traditional shapes, but mechanized production cannot cut slate as thin as the nineteenth-century hand-cut process. This may cause some dimensional problems with existing flashing and intersections with other materials (Figure 7-33).

Slate is an enduring roof material because it is impervious to weather and is fireproof. It does crack, however, and it is not entirely maintenance free. Slate is available in several qualities. Minerals in slate impart a variety of colors, which makes it difficult to match slates when parts of a roof are replaced. To overcome this problem, a highly visible nonmatching replacement can be exchanged for an original slate in an obscure part of the roof.

Cracked slate can be repaired by working roofing cement into the crack and cleaning away the excess cement so that it does not show (Figure 7-34). To remove a damaged slate, cut the nails holding it to the sheathing or furring strip with a hacksaw or nail cutter. Trim the new slate to size and install it with copper- or zinc-coated nails. Sometimes a strip of copper is nailed where the slate is to be installed, then bent up over the bottom edge to hold the slate in place. All nailheads must be coated with roofing cement.

Figure 7-33 Old slate roof, City Hall, Oswego, New York. Prior to restoration, the deteriorated old nineteenth-century slate roof had been repaired, losing its original pattern and the iron roof cresting.

Figure 7-34 Restored slate roof, City Hall, Oswego, New York. The restoration of the mansard roof included reproduction of the original colored slates set in a characteristic nineteenth-century geometric pattern. The copper flashing and the iron cresting have been restored as well. Because of the bold simplicity of the building's stone facades, restoring the detailing and color of the mansard roof was of particular significance.

Metal Shingles

Metal shingles or tiles were available in the United States beginning in the 1870s, but declined in popularity in the 1920s. They were replaced by asphalt shingles, which were plain and did not require painting every three to five years. Metal shingles were easily stamped with decorations or patterns that could be painted to simulate wood, slate, or terra cotta.

Primarily made of tin and terneplate, sheet iron, and galvanized iron, metal shingles provided an inexpensive, lightweight, and fireproof roof. To restore metal shingles, follow the advice under "Metal Roofing," later in this chapter.

Asphalt Shingles

Asphalt shingles date from the late 1800s and are still in use today. They have a life span of about 30 years, which is comparable to wood shingles. Early asphalt shingles were fabricated in interlocking and geometric shapes, many of which are no longer manufactured (see Figure 7-35). After World War II, asphalt shingles were made to duplicate traditional slate and wood shingle materials.

Figure 7-35 Old asphalt shingles, Camp Sagamore, Adirondacks, New York. Now the most common roofing material, asphalt shingles were originally used because they were cheap and were often applied over leaking old wood shingled roofs.

Asphalt tiles are made with felt that is impregnated with asphalt and covered on the exterior surface with mineral granules. When the granules are worn off, the roofing needs replacement. For a historic restoration, this may be the time to research the original roofing material and color.

Although asphalt shingles are cheaper than wood shingles and have a better fire-safety rating, they cannot successfully imitate the shadowline and texture of wood shingles. If asphalt shingles are to be used again, consider using a heavier weight in order to prolong the life of the new roof.

Asphalt shingles are made in several weights. In roof patches, the weight of the replacement shingles should match the originals. They can be replaced using a process similar to wood shingle replacement, since they too are laid in overlapping courses. Problems arise when asphalt shingles have been installed over wood shingles that were probably wet and deteriorated when they were initially covered. The asphalt covering prevents air from circulating to the wood shingles, thereby accelerating the deterioration. When this occurs there is no option but to remove both roofing materials and start anew.

Asphalt shingles tend to curl, but this is easy to fix. Roofing cement is spread under the raised part and the shingle is weighted down with a sandbag until the cement sets. Torn shingles should be cemented and then nailed along both sides of the tear. The nailheads are, of course, covered with roofing cement. If the tear has widened, a piece of copper several inches wider than the tear should be cemented into place under the tear and then nailed through the shingle.

Figure 7-36 Stanford University, Palo Alto, California. The original campus buildings of Stanford University eschewed the Gothic Revival of the eastern ivy league in favor of the adobe tradition of the local Spanish missions. The focal structures of the Beaux-Arts campus plan have clay tile roofs, frescoed stucco walls, and wrought-iron detailing.

Clay Tiles

In the former Spanish territories in the Southwest, unglazed terra cotta tiles were the typical roofing material for adobe buildings (Figure 7-36). These roofs are fabricated from a few basic shapes, with no special

accessories for ridges, caps, or valleys. These roofs can last for centuries in the hot, dry climate of the Southwest. Unglazed terra cotta roofs on Mission Revival style buildings in colder and more humid areas do not fare as well and, therefore, a glazed Mission roofing tile was produced in the late nineteenth century. Many of the elaborate Richardsonian Romanesque and Queen Anne style buildings constructed at the turn of the century were roofed with glazed Mission tiles. The complex roof shapes, oriels, dormers, conical turrets, and bay windows required complex glazed-tile accessories to form hips, ridges, caps, finials, and other trim pieces (see Figure 7-37). It is almost impossible to find matching replacements for missing or damaged glazed terra cotta elements, but new glazed terra cotta is now being custom manufactured.

Clay tiles, also called ceramic tiles, are flat or curved. The latter are frequently seen on Mediterranean-style buildings. Both are installed on furring strips nailed to wood sheathing, with the tiles overlapping in courses. Because the tiles interlock and each successive course is overlapping, they are not easily removed and replaced in the same way as other roofing materials. Again, roofing cement is applied under the tiles and over the nailheads.

Figure 7-37 Glazed tile roof, Brooklyn, New York. Heavy glazed tile roofs have many individual components and special shapes which form valleys, ridges, caps, finials, and tapered elements for cornical towers. Because they are interlocking it is difficult to replace damaged tiles; finding a replacement to match faded glazes is also challenging.

Flat Roofs

Flat roofs are covered with built-up roofing that usually lasts up to 30 years. The roof is installed in layers of asphalt-impregnated roofing felt and mopped with tar or asphalt. The final coat is covered with gravel or marble chips to reflect the sun and prevent cracking. Built-up roofs are susceptible to cracking and blistering, which can be patched easily.

Cracks that are less than 2 inches wide can be filled with roofing cement. With larger cracks, you must cut around the damaged area and fill it with a piece of roofing felt, leaving a 2-inch overlap all around. Nail the patch down and cover it with roofing cement. Then cover the patch with a larger piece of felt extending well beyond the first patch and nail and cement it into place. The final cement coating should be covered with fine gravel.

Blisters must be cut open with a knife to make the old felt lie flat. Pour roofing cement under the area and nail down the edges of the cut. Follow the same procedure as for cracks, and install a large patch.

Metal Roofing

In rolled sheets, metals such as lead are ideal for large areas of roofing. Sheets can be heat-sealed together to create waterproof joints. In the eighteenth and nineteenth centuries, lead and copper were the only flexible materials available for covering undulating roof surfaces such as domes and spires. Smaller flat sheets were used for flashing on wood, slate, and clay tile-roofs at intersecting roof slopes, chimneys, parapet walls, gutters, and cornices. Metals were also pressed into shapes resembling tiles.

Lead Roofing. Lead resists most types of corrosion but is susceptible to abrasion and creep. If deterioration is widespread, the lead may have to be replaced with a hard alloy that will resist the causes of mechanical failure. Small, localized failures can be treated by inserting new lead panels into holes and melting their edges into the original roofing.

When the deterioration is not severe enough to warrant replacement, the surface can be cleaned and painted or coated with a layer of copper. The recommended cleaner for corroded lead surfaces is Versene powder or acid. It removes the crust and ensures that the surface is free of formic and acetic acids.

If the roof understructure needs repairing, the lead sheets can be cut free, rolled up, and put back later. However, if the lead has become uneven from age and corrosion, it can be melted and recast into new sheets of uniform thickness. Before replacing any lead sheets, cover the roof's sheathing with rosin paper, so that the lead will move freely with thermal expansion. At the same time, provide expansion joints at a maximum of 9 feet and 24 feet centers in each direction.

Figure 7-38 Old terne roofs, Camp Sagamore, Adirondacks, New York. If properly painted, old terne plate roofs can remain in service for more than a century under the most severe weather conditions.

Tin and Terneplate. Tin is found in roofing as a constituent of terneplate. Terneplate is iron or steel coated with a mixture of lead and tin. Plates coated only with tin, as in pressed ceiling panels, are called tinplate. When tin is exposed to the atmosphere, its oxide forms a protective coating. But when the tin coating is damaged, the iron or steel underneath will corrode because of the atmospheric and galvanic action between the tin and the iron (Figure 7-38).

Most terneplate roofs respond to localized repairs and repainting. The most common repairs are to opened seams and popped-up nailheads. Both should be cleaned and resoldered using 50 percent pig lead and 50 percent block tin. Sheets should be refastened with galvanized iron or steel nails, or with tinplate cleats.

If damaged sections are removed and replaced with new pieces, the same material must be used. If the material is not the same, galvanic action will destroy the repair. Advances in metallurgy have improved traditional terneplate, and it is now usually made with terne-coated stainless steel or replaced with lead-coated copper. Use these materials if the whole roof or siding is to be replaced. An advantage of these new materials is that they do not have to be painted; a disadvantage is that stainless steel is extremely hard and therefore difficult to bend and cut.

Painting original terne roofs is mandatory and painting both sides of the sheets is preferred. Use primers containing linseed oil with red lead or iron oxide. Two coats of an oil-based, high-gloss paint finish should be applied two weeks apart. This paint must be compatible with the primer.

Flashing, Gutters, and Downspouts

Leaks are often caused by deteriorated flashing rather than the roofing material itself. So before ripping into the roof, be sure to inspect the integrity of the flashing. The purpose of flashing is to bridge the junction between two surfaces of roof materials. Flashing materials include aluminum, lead, copper, tin, galvanized iron, lead-coated copper, stainless steel, and terne.

Small holes in metal flashing can be patched by soldering, or by priming the area and then filling it with roofing cement. Larger holes are patched with the same type of metal. Cut the patch one inch larger all around, fill the hole with roofing cement, and push the metal patch into it. Loose flashing can be nailed back into place with nails of the same metal. Cover the nailheads with roofing cement. Flashing that has pulled out of masonry joints can be sealed back into the joint with two coats of roofing cement. The first goes between the flashing and the brickwork; the second is liberally applied after the flashing has been set in place.

Since the objective of a good roof is to carry away water as quickly as possible, gutters and downspouts must be in top condition. If they are undersized, incorrectly pitched, or defective, water will get under the lower courses and quickly deteriorate the roofing material. Restoring gutters and downspouts often requires no more than simple maintenance and regular cleaning (see Figures 7-39 and 7-40). The metal needs repainting, and holes and joints should be filled and caulked. Rust spots should be wire-brushed, primed with a rust-inhibiting primer, and finished with two coats of paint.

If a section has to be replaced, be sure that the new piece is of the same metal as the old sections, usually aluminum or galvanized steel, so that there will be no galvanic deterioration. The same prescription applies to installing new brackets for gutters and downspouts.

PAINT

It is taken for granted that wood surfaces in buildings are treated with paint, stain, or preservative. But it wasn't always so (Figure 7-41). Seventeenth-century and early eighteenth-century buildings were seldom painted, because paint was scarce and therefore expensive. In most small communities only public buildings, such as churches and courthouses, were painted. Oil-based paint was prepared by painters who ground pigments and white lead together in linseed oil. In the middle of the nineteenth century, health authorities became aware of lead poisoning, and zinc oxide was substituted for white lead.

Figure 7-39 Repairing gutter, mansard roof, Brooklyn, New York. Restoring an 1870s slate-covered mansard involves work not only on the slate roofing material and copper flashing, but also on the wood framing beneath that has rotted due to water seepage.

Figure 7-40 Gutter detail, mansard roof, Brooklyn, New York. Failure of the metal-lined gutter has led to the deterioration of portions of the gutter, the cornice, and windowsill of the dormer, which must be replaced.

Figure 7-41 Buttolph-Williams House, Wethersfield, Connecticut. During the early colonial period wood was plentiful and paint was costly. Many ordinary structures, such as houses and farm buildings, were left unpainted.

Driers, such as metallic salts, were added to paint to start the oxidation that leads to drying, and turpentine and mineral spirits were added to thin the paint without affecting the drying of the surface. Grinding and mixing pigments and bases into oil and thinners by hand created variations in color and texture. The first ready-mixed paints were not marketed until just before the Civil War, and it was a long time before they became widely available.

Whitewash, a mixture of lime and other ingredients dissolved in water, was, until the late nineteenth century, the most widely used house paint. But it didn't protect the wood for very long and it needed to be renewed each year. In rural areas, lime and other ingredients were mixed in milk, producing a coating slightly more durable than whitewash. This coating was sometimes called buttermilk paint.

When Is Repainting Necessary?

The urge to repaint is strong, but painting is sometimes an unnecessary effort and expense. If exterior paint on wood shows no signs of chalking, blistering, peeling, or cracking, there is no need to repaint. Nor, says the Technical Preservation Services of the National Park Service, should you paint merely because the old colors have faded.

The reason for the caveat is that when coats of paint build up, the resulting thick coating is less resistant to thermal stresses, and the underlying coats are less able to withstand the shrinkage of a new layer as it dries. If you repaint, you may be adding the layer that triggers the shrinkage causing the oldest layers next to the wood to lose adhesion.

Eventually, paint build-up causes a loss of detail on wood-carved decorative elements. Admittedly, it is not always possible to ascertain when the critical thickness has been reached, so when in doubt clean the siding instead of painting it. If a change in color is then wanted, new trim colors may be an acceptable substitute.

Figure 7-42 Chalking paint, Wethersfield, Connecticut. Too much chalking may eventually disintegrate the paint film that protects the wood, leaving it exposed to the weather.

Exterior Paint Deterioration

In order to restore the sharp detail lost in paint built up over the years, an old building will eventually require paint removal down to the bare wood. But to avoid adding unnecessary coats of paint, it is important to ascertain whether the deteriorated surface can be corrected. Some conditions that do not require paint removal or repainting are dirt, mildew, chalking, and stains.

Dirt. Dirt, soot, and environmental grime build up on painted surfaces, and if they are not removed before repainting, they will prevent good adhesion and cause peeling. Most urban grime can be washed off with a garden hose, a soft bristle brush, and household detergent. If cleaned well, the surface may not need repainting.

Mildew. Mildew thrives in damp, dark areas. To distinguish mildew from dirt, put a drop of bleach on it. Mildew will turn white; dirt remains dirt-colored. Mildew can be washed off with a solution containing bleach and nonammoniated detergent. However, it will reappear unless the environmental conditions that fostered it are altered. When repainting such areas, use specially formulated mildew-resistant primer and finish coats.

Chalking. Chalking, or powdering, occurs because the resin in the paint disintegrates. A little chalking is good because rainwater rinses off the chalk and carries away dirt, thus providing an ideal surface for repainting. However, too much chalking will wash onto other surfaces and mar them; it will also disintegrate the paint film that protects the wood. A household detergent applied with a soft brush will remove chalking. Thoroughly hose off the detergent and allow the wood to dry before applying a nonchalking paint (Figure 7-42).

Stains. Stains from rusting nails or metal elements can be stopped by hand-sanding the metal and coating it with rust-inhibiting primer and two finish coats. Nails, wherever possible, should be countersunk and spot-primed, and the holes should be filled with high quality wood filler. Stains from the natural extractives of some woods, such as red cedar and redwood, can be cleaned with a solution of equal parts of denatured alcohol and water, and then dried and coated with stain-blocking primer.

Limited Paint Removal

Some deteriorated surfaces can be corrected with only limited paint removal, thus avoiding the tedious and costly process of stripping paint to the bare wood. Some of the conditions requiring limited paint removal are crazing, intercoat peeling, salts, blistering, and wrinkles. Peeling to the bare wood and cracking necessitate complete paint removal.

Crazing. Crazing results from several layers of paint becoming too brittle to expand and contract with temperature changes. The crazing may be so fine that it is difficult to detect, but it will lead to cracking when moisture enters the cracks. The surface should be sanded before repainting.

Intercoat Peeling. Intercoat peeling is often found under eaves and on porches, where rain has not washed away salts from airborne pollutants (see Figure 7-43). If the surface had been washed properly, peeling would have not occurred. Incompatible types of paint also cause intercoat peeling. Oil paint over latex paint will sometimes peel because, after aging, oil paint is less elastic than the latex. Conversely, latex paint will peel off old, chalking oil paint because it cannot penetrate the surface to form a bond.

Figure 7-43 Intercoat peeling reveals varnished doors, Brooklyn, New York. Incompatible types of paint can cause intercoat peeling. Peeling is also found under eaves and porches where rain has not washed away salts from airborne pollutants.

Salts. Salts can be washed off and the dried surface can be sanded and repainted. With incompatible paints, the top coat should be scraped, sanded, and coated with an oil-type exterior primer. An oil or a latex finish coat can be applied.

Blistering. Solvent blistering results from solvent-rich paint being applied in direct sunlight. The top surface dries too quickly and the solvents trapped beneath the dried film create blisters. When a solvent blister is cut open, another paint layer is visible, but when a moisture blister is opened, it reveals bare wood. The solvent blister can be scraped, sanded to the next sound layer, and then painted. But it should not, of course, be worked on in direct sunlight.

Wrinkles. Paint wrinkles when (1) a second coat is applied before the first coat dries, (2) it is not brushed out sufficiently, or (3) when the ambient temperature is too high at the time of application. Wrinkles are removed by scraping and sanding. Repainting procedures should follow the manufacturer's instructions.

Complete Paint Removal

Sometimes the condition of the paint is so poor that nothing will adhere to it, and consequently the surfaces must be stripped bare before any new coats are applied. Two conditions requiring total paint removal are peeling and cracking.

Peeling. Peeling to bare wood is caused by moisture behind the paint film that is preventing adhesion. The damaged paint can easily be scraped and sanded, but that will not solve the problem. First, the cause of the moisture has to be removed from exterior and interior sources. Flashing, gutters, caulking, and so on, should be examined and repaired, and, if necessary, exhaust vents and fans should be installed to disperse interior moisture build up.

Cracking. Cracking and scaling develop from crazed spots where water has seeped in. If only the top layers are affected, they can probably be scraped and sanded to a sound layer. If not, the paint has to be removed to bare wood with thermal or chemical methods. Bare wood should be primed within 48 hours, then repainted.

Safe Methods for Removing Paint

Always use the gentlest possible method for removing paint from a particular element. Taking short cuts in removing paint from old buildings can cause serious and irreversible damage. The three methods to choose from are (1) abrasion, (2) thermal, and (3) chemical.

1. Abrasion entails scraping with putty knives and paint scrapers. Putty knives are pushed under the paint, working from a loose area into an undamaged area. Scrapers have replaceable blades and are pulled in a raking motion into the damaged area of paint.

Sanding should be done by hand wherever possible. Orbital sanders are preferred to belt sanders when power tools must be used. Either way, use coarse-grit, open-coat flint sandpaper because it clogs less quickly and is less expensive than other types. Follow the grain when sanding, and for moldings use a sanding sponge to reach into the grooves and irregularities. Rotary disc sanders and wire strippers are not recommended because they leave visible marks and can destroy wood surfaces. High-pressure hoses (over 600 psi) are not recommended because they force water into the wood. Sandblasting should never be used on wood because it destroys the surface and erodes moldings and other details.

2. To completely remove paint from woodwork, two thermal tools are recommended: electric heat plates and electric heat guns. Blowtorches should not be allowed near a preservation project.

Heat plates held close to paint soften the layers without vaporizing lead paint. When the paint softens and blisters, it is scraped off with a putty knife. With practice, the heat plate can be used in continuous motion, thus lessening the risk of scorching the wood.

Heat guns work at about the same temperatures as heat plates—50°F to 80°F—and use the same 15 amps of power. The gun blows hot air onto the paint, which blisters it so that it can be scraped off. Both types of heat work best on thick layers of paint since there is less chance of scorching the wood.

The gun is particularly effective on details because the nozzle can be directed where the operator wants it. The heat plate is best suited to flat surfaces, and both tools are necessary for most paint-stripping situations.

3. Chemical removal is used for windows (a heat gun cracks glass), for details that a gun could not soften, for varnish found on elements after the paint has been removed by heat, and for detachable elements such as shutters and doors.

Chemical solvent-based strippers are formulated as liquids or, with the addition of a thickener, are available in semipaste form. The latter is used for vertical surfaces. Both types create toxic vapors, so the work space must be well ventilated. For both exterior and interior work, a respirator with a filter for organic solvents should be worn. Also, the strippers should be kept off the skin.

Water-based removers, although convenient, ultimately create more problems than their convenience is worth. Although they can be rinsed off with water, they often leave a gummy residue on the wood that requires more solvents to remove it. Furthermore, they can raise the grain of the wood. The U.S. Department of the Interior's Office of Technical Preservation Services recommends regular strippers over water-rinsable strippers.

Figure 7-44 Wilderstein, Rhinecliff, New York. The proverbial ghostly Charles Adams mansion with peeling paint is a preservationist's nightmare requiring an enormous rescue effort. The paint may indeed have faded so completely that it can no longer serve as a reference for color matching. More serious is the fact that the wood elements are no longer shielded from weather exposure. There is a point where the deterioration caused by deferred maintenance cannot be reversed.

Cleaning and Painting Metal Surfaces

Metal surfaces have to be cleaned thoroughly with an abrasive process before painting. The degree of cleaning varies from blasting, which exposes flecks of clean metal, to complete removal of all paint and contaminants. How much should be removed depends on the state of deterioration, how much adhesion is required, and the budget for labor. (See the section on Rehabilitation, Replacement, and Maintenance of Specific Materials or Parts in Bibliography.) If the metal is aluminum, steam cleaning may be possible. All cleaned metal surfaces should be washed with a coat of cold phosphoric acid and then repainted as soon as possible to prevent further oxidation.

Matching Old Paint Colors

If the color and texture of paint is to be faithfully restored, the pigments that were available at the time of the building's construction should be investigated and the aging of oil paints, which may have faded, should be taken into account (Figure 7-44). Even if the restoration is not taken to such purist lengths, some thought should be given to the colors that were originally available. Early builders were adventurous with color. They often used striking tonal combinations, and they also used subtle shading of hues of the same color. Recent facsimile editions of nineteenth-century sample books illustrate the oil paint schemes with color chips, and these can be matched with contemporary paints from several major paint manufacturers.

Selecting New Paint

The original colors of a building can be discovered by cutting out a section of the surface paint with a razor blade, all the way to the bare wood, and examining the successive layers through a magnifying glass. If the original paint is the topmost layer, a patch should be cleaned with solvent to reveal the color without grime. Old coats will have faded, so make allowance when matching new paints against the old (see Figure 7-45).

Figure 7-45 "Painted Ladies," San Francisco, California. Much of the nineteenth century was more colorful than we think. Several manufacturers have reproduced fascimile paint catalogs and offer an authentic range of period colors. Some preservationists get carried away with the vast selection available and find it hard to limit themselves.

Because historic buildings were painted with oil-based paints, oil paint should be used when adding another coat. Oil paint shrinks less than latex, and oil paint will be less likely to pull the original coats loose. Another reason for reapplying oil paint is that is has a better chance of penetrating any chalky residue left on an old coat that has not been cleaned thoroughly.

If latex paint is preferred for the final coat, it should be preceded by an oil primer to create a flat, porous surface. When all the original paint has been removed, an oil primer should be applied; this may be followed by either an oil-type or a latex top coat. In both cases, the primer and top coat should be from the same manufacturer, and top coats should be applied as soon as the primer dries.

Reproducing Old Painting Methods

Painters of the past used paints made with hand-ground pigments that were less uniform than today's factory-blended, off-the shelf paints. A fine point of restoration, particularly when purity is a prime objective, is to re-create the look of old paint by using old paint-mixing techniques. In buildings where the original surfaces were whitewashed, however, be aware that the price for replacing subsequently added paint layers with whitewashing will be high annual maintenance costs.

Techniques for graining, marbling, and other *faux* processes, such as imitation masonry on building exteriors, were perfected over the course of the nineteenth century (Figure 7-46). Wood was disguised as masonry by adding sand to paint or by working sand into tacky paint after application. *Faux* masonry appears in architectural construction and decorative details such as flush boarding, quoins, cornices, and rustication. A number of imitative techniques are discussed in the section on interior finishes (see Chapter 8).

PLASTER

Except for log cabins, almost all American buildings constructed between the colonial period and World War II have plaster interior finishes. Since 1945, traditional wet plaster has been all but supplanted by drywall, which is faster, simpler, and less expensive to install.

Traditional plastering needs considerable preparation and is quite messy. It also requires periods of waiting time for setting and drying, which disrupts other processes and prolongs the completion process. Plastering is a dying skilled trade. There aren't many plasterers available even to restore walls and ceilings, which is lamentable since major repairs to plaster cannot be done by amateurs. The availability of a skilled plasterer to restore traditional decorative plasterwork is likely to be a major factor when deciding whether to restore a building's interior with plaster or drywall.

Figure 7-46 Redwood Library, Newport, Rhode Island. Many eigtheenth-century American buildings constructed of wood were modeled on classical European prototypes constructed of masonry. Sometimes sanded paint was applied to imitate stone texture.

Figure 7-47 Adobe maintenance, San Diego, California. One of the serious conservation problems of old adobe buildings is their vulnerability to fungal growths. Often dampness in the wooden roof pole framing is the source of the problem. Periodic spraying with fungicides will help to control this problem.

Figure 7-48 Stucco mansion, East Hampton, New York. Many American historic revival buildings imitate the appearance of exotic structures from much more temperate locales. Stucco does not fare well in severe northern winters, particularly when it has been applied over wood framing which is subject to seasonal expansion and contraction.

Traditional Methods of Plastering

The basic composition of plaster has not changed since Egyptian times. It is still made from gypsum rock and is still troweled onto walls in two or three coats, starting with a coarse mixture called a brown, or scratch, coat and finishing with a fine white coating. The scratch coat contains gypsum plaster mixed with sand and vermiculite or perlite as a binder. (Before the nineteenth century, animal hair was used to bind plaster.) A second coat of similar constituents is applied to the scratched surface of the base coat. Finishing coats contain a type of plaster, called gauging plaster, that is mixed with lime to create a smooth chalk-white surface. Early plastering techniques vary by region but have in common relatively soft plaster applied over hand-split wood lath. Chair rails made of wood were often applied to prevent furniture from marring the plaster surfaces. Later, lath was replaced by expanded metal as the support material.

Generally walls had hollow spaces between the plastered interior surfaces and the exterior walls of wood clapboard or shingles. Some primitive forms of insulation were provided by clay, seaweed, or crude bricks being packed into the spaces between the rough timber framing. The concept of vapor barriers and thermal insulation did not exist. The sudden introduction of modern storm windows and insulation into an old building can totally unbalance the equilibrium of the various old materials. Neither excessive condensation nor dryness is good for the wood framing or lath that supports plaster. Excessive movement or shrinkage in the structure, which modern climatization-measures promote, can cause cracking and deterioration of plaster finishes.

Mud Plaster. In some parts of the country, earth is used as a construction material. Sod, mud bricks, and adobe have produced a distinctive vernacular architecture. Mud-plastered walls were usually coated with lime whitewash. Specialized chemical treatments must be periodically applied to old earthen buildings to prevent the growth of fungi (see Figure 7-47).

Stucco. Stucco is a form of cement plaster that is water resistant and can be used on exteriors. Stucco and cement plaster are types of masonry finish that are softer than brick or terra cotta. As described in the masonry section (see pp. 106–113), these materials were usually applied over porous rubble-stone or brick walls to discourage moisture penetration (see Figure 7-48). In many historic revival styles, especially those derived from Mediterranean origins, stucco-covered masonry was imitated on wood-framed structures. A support system of wire lath was applied over wood battens nailed to the exterior wood sheathing. Walls were often studded with decorative glazed clay tiles and capped with heavy Mission tile roofs.

In northern climates with extreme freeze-thaw cycles, stucco does not weather well and is subject to severe cracks and water penetration. Similar problems occur in mock Tudor structures with half-timbered gables and bays (see Figure 7-49). The application of masonry coatings and paints will sometimes complicate finding the source of water penetration because they conceal the cracks. Some structures combine stucco applied over wire lath with stucco applied directly on masonry elements such as chimneys, porticoes, and foundation walls. The various rates of moisture absorption cause cracks at the junctures of the brick and wood portions of the structure because of differing expansion and contraction properties (see Figure 7-50).

Plaster Applied Directly to Masonry. Sometimes plaster is directly applied to masonry walls without lath (see Figure 7-51). This is often done on chimney breasts and party walls in row houses. However, builders recognized that this practice was not successful on exterior walls because of dampness and condensation.

Figure 7-49 Mock Tudor residence, Riverdale, New York. In the 1920s and 1930s, prosperous suburbs sprouted mock Tudor mansions, with deliberately sagging slate roofs, diamond-paned leaded casements, and many "authentic" details.

Figure 7-50 Column refacing, Chenango County Courthouse, Norwich, New York. The deteriorated cement stucco rendering of the columns is refaced to restore the sharpness of the detailing of the classical flutes. When the repaired masonry is not to be painted, matching stone cut from deteriorated portions can be pulverized and combined with the finish cement stucco coat in the areas to be patched. This provides a better and more stable color match than masonry dyes, which are subject to fading and do not reproduce the original texture.

Figure 7-51 Art Deco library, Berkeley, California. An elaborate Art Deco fresco has been carved and painted into the exterior stucco finish of this California public library.

Plaster on Lath. The hollow void between the plaster on lath and the exterior wall provides an excellent place to conceal new wiring and plumbing. Before such placement is decided on, test holes should be made to determine whether hollow spaces exist. However, chopping into walls that have been directly plastered is risky and may cause unanticipated damage to other material and decorative plaster.

Repairing Bulges

If there are bulges in a wall, they may be cured by screwing the laths back to the studs. This is done by drilling through the plaster and inserting a screw with a washer into a countersunk hole. When the bulge is remedied, the countersunk heads should be plastered. If the wood laths were intact beneath the bulging plaster, the same screw system may secure the plaster back against the laths. If the wood lath in the damaged area is missing or loose, it can be replaced by modern wire mesh, but this must be shimmed to make up for the difference of thickness. If the plaster is filled to the full depth it may fall out because of its own weight.

Usually, bulging plaster in small areas needs to be chipped out and the hole filled with a plaster patching compound. If the hole is about three inches in diameter, plaster should be substituted for the compound. The repair of damaged areas should proceed in the same way that the wall or ceiling was originally plastered; that is, it should be built up in two coats. The base coat should have a fine aggregate in it, such as sand or vermiculite, and should fill half the depth of the hole. Then, scratch a diamond pattern on the surface before it sets; mix the finish coat to a smooth consistency and trowel it into the hole and beyond its edges. Finally, with a steel edge, smooth the plaster flush with the wall and do not sand it until the new coating is hard.

Repairing Cracks

Plaster walls sometimes crack when a building moves. These are most often seen as diagonal cracks that start at the top of window and door frames. If the cracks are patched carelessly, and the building moves again, the patch will open up. To avoid an endless patching cycle, the crack has to be filled and then covered with fiberglass tape, which, in turn, is coated with flexible patching compound that is feathered into the surrounding plaster. When the area is sanded and painted, the tape will not show, and the crack will not reopen.

Plaster walls and ceilings that are basically sound, but have surfaces marred by crazing, can be papered or covered with a fine scrimcloth and then painted. Another, more drastic, solution is to scratch (scarify is the technical word) the entire surface so that it will be rough enough to receive a new finish coat of plaster.

Because wet plaster is built up in a series of hand-troweled coats, its thickness varies. (Plasterboard, which is a manufactured product, is absolutely even and easily detected.) Thus, most attempts to patch uneven old plaster with drywall are unsuccessful. Also, most modern drywall products are not as thick as the layers of built-up plaster, which means that patched areas must be shimmed to achieve a flush finish. Spackling compound, which is commonly used in the taping of drywall, is sometimes used for minor patching, but it does not have the same characteristics as plaster and does not always adhere permanently. Extensive patching with spackling compound will not yield satisfactory results and may cause serious problems if it becomes loose after elaborate paint finishes or wall coverings have been applied. Although the existing plaster finishes in an old building may be in reasonably good condition, installation of new plumbing, heating, and wiring may necessitate a considerable amount of patching.

Repairing Ceilings

One of the most common repairs in old buildings is to cracked or deteriorated ceilings. Ruined ceilings are often caused by roof or plumbing leaks in which dry old plaster has soaked up moisture, gotten heavy, and stained the wood lathing, causing it to pull away from the wood framing or rafters. In severe instances, the old lathing nails become so rusted that they fall out and the entire lath-and-plaster ceiling collapses.

When there is extensive interior demolition in an old building, the adjacent rooms must be checked for cracks, bulges, or signs of deterioration in the old plaster finishes. By cautiously tapping old plaster surfaces, an experienced mechanic can ascertain which portions have lost their bond and require patching, repair, or replacement.

If an entire ceiling must be removed it may be easier and cheaper to replace it with plasterboard. The old lathing and plaster must be carefully removed so that there is a minimum of damage to the walls where they intersect the ceiling. If there is a cove or curved detail above the picture molding, the removal of the flat central portion of the ceiling becomes a risky operation. Matching missing or damaged portions of the cove is much more difficult than the flat portions and may require the creation of a template to match the existing profiles. The reproduction of existing profiles may be unavoidable anyway because of changes in layout, or the construction of new partitions that intersect old coved plaster walls.

Since reproducing cove profiles is a nuisance, it may be better to avoid the risk of damaging them. One way to do this is by removing the wood baseboard trim and running new electrical circuitry horizontally. However, unlike the modern technique of applying wood baseboards and trim after the completion of the drywall, in the traditional method, the wood "grounds" for the trim were set before the plastering, a condition that complicates its removal. Despite the risk of some minor damage to the adjacent plaster surfaces, the patching of flat vertical surfaces is much easier than the replacement of ceilings, coves, and molded plaster decorative elements.

Molded Plaster

Decorative plasterwork that was originally cast in molds can be replicated by taking a mold from the existing plasterwork. For example, a broken ceiling medallion can be patched by taking a mold, making a copy of the broken section, and installing it back on the medallion. Of course, there is a lot of skill needed to cut out the broken piece and hide the transition between the new piece and the original. Many of the original, heavily carved plaster decorations with lush three-dimensional scrolls, leaves, garlands, and trophies are actually composites cast in separate pieces and assembled in place. The old plaster ceiling must be thoroughly checked to make sure that the lath and plaster are securely attached to the structural framing before attaching newly cast reproductions of decorative elements. Some reproduction moldings and medallions are now available in lighter weight synthetic materials that can be painted to match the adjacent plaster finishes. The weight factor may be significant if the heavy old plaster ceiling is in poor condition. Also, some molded plastic decorative reproductions may not satisfy fire-code requirements for public places.

Applying precast plaster decorative elements seems like child's play compared with reproducing run plaster moldings. These moldings, which extend around the perimeter of a ceiling or are in the frieze of a wall, are, in effect, extruded in place. This means that wet plaster is gradually built up on the wall or ceiling and a template of the decorative section is pushed back and forth by hand along a trolleylike device until the desired profile is achieved. It is a two-handed job that requires great dexterity working with the rapidly setting plaster and a keen eye to compensate for the irregularities of old walls.

Substitutions can be made. If only small sections of a molding is broken or missing, it can be copied from the original section on a bench and then, using wet plaster, pasted into position. Again, the new piece will have to be eased into the original so that no joints or changes in profile show.

The process of taking molds and making castings is explained in many how-to-do-it manuals for artists and sculptors. A local sculpture studio may be helpful in making molds and castings of decorative architectural elements if skilled plasterers are not available. Some specialized aspects of decorative and ornamental plaster are discussed in Chapter 8.

WATER SUPPLY AND PLUMBING

Rural and urban projects have differing water supply considerations. Rural properties may contain their own independent water supply, and one of the first on-site investigations should be of water quality and availability. If the quality is unsatisfactory, it may be necessary to provide water treatment equipment.

An old water source may not be sufficient for current or future needs. With an isolated property, the availability of an adequate fire reserve may also be significant. The need for a water storage tank, particularly if pressure is low and the source must be elevated, may indicate a concealment problem in order to maintain the original appearance of the property.

In urban areas, where the ubiquitous water tank is a traditional feature of the skyline, it is possible to replace these tanks with pumps in the cellar. Removing a heavy water tank can relieve certain structural problems as well as provide additional usable space on rooftop areas.

Water Supply Piping

Some nineteenth-century water supply piping was made of lead and should be replaced since lead is now known to be a serious health hazard. Cast iron was more commonly used for water supply and waste lines, but the minerals in the water contributed to corrosion of these pipes. Starting about 80 years ago, water pipes were made with galvanized steel. Later, copper pipes became standard for water, but the old cast-iron waste pipes often remain. Waste pipes resist deterioration because they are not subject to the high pressure of water supply lines, and detergents prevent build-up of deposits.

Because their remaining life will be short, cast-iron water pipes, whatever their condition, should be replaced. Galvanized pipes should be inspected by disconnecting the pipe at a fixture and looking for deposit build-up. The rule of thumb is that if half the diameter is filled, the pipes need to be replaced. It is tempting to economize by retaining old plumbing if it is still functioning, but this is shortsighted. If a major renovation is planned, the pipes should be changed because it will be even more expensive to do after the new work is completed. The plumbing work should be entrusted to a licensed plumber regardless of the size of the job. The plumber should be familiar with the local conditions, such as water pressure and minerals, that cause pipes to corrode or clog up.

The vents and traps on the waste lines may not meet present code requirements. Convert them immediately, since there is always a danger of sewer gas explosions. If previous renovation work was done by amateurs, make sure that plastic pipes have not been illegally installed and make sure that galvanized pipe is not joined directly to copper pipe (use an electrolytic union, often required by codes, to reduce the galvanic action between the two metals).

New Plumbing

When new bathrooms, kitchens, or other water-supplied rooms are added to a building, their location should be chosen to minimize the plumbing pipe runs. If a new room can be added over a former bathroom or kitchen, there will be space for the vertical pipes through the floors, and the supply and waste pipes entering the building will usually be close by. If plumbing is to be rerouted, avoid concealment behind decorative plaster finishes. In order to assure adequate access for repair, also provide local shut-off valves.

Waste Disposal

Rural buildings not on a municipal sewage system generally have their own septic tanks. There are specialists who service septic systems, and before replumbing a building can proceed, the condition of the septic tank must be assessed. The tank may only need to be pumped out. However, if the walls of the tank are in poor shape, and if the waste water from the building merely trickles into the tank, the system must be renewed. Even if the old system is still functioning it may not be adequate to the new needs of the restored building. Old residential installations may not have the capacity to serve a dishwasher or clothes washer and additional bathrooms. Converting former lofts or institutional buildings for residential or commercial use may also require new waste disposal systems.

Environmental regulations are becoming increasingly strict, and even a functional old system may not fulfill modern requirements. The problem of waste disposal may be critical to the selection of an adaptive use, particularly on waterfront sites.

Reusing Old Plumbing Fixtures

Some restorers love old plumbing fixtures, but outside of private homes, elegant old sinks and bathtubs often have to be replaced to conform to current building codes. If the occupancy of the building allows old fixtures to be used, the existing equipment can be cleaned and repaired. If the old fixtures are to be reused, they must be carefully removed and protected during construction. Old porcelain may need refurbishing, and the sink and tub probably will have to be taken to a specialty firm. Chipped porcelain enamel can be restored; cracked glazed ceramic fixtures cannot. The use of touch-up paints is not recommended for porcelain because they do not adhere well and seldom match.

It is the plumbing, not the fixtures, that creates the major problems since old pipes may need replacing and are often buried in walls. Also, pipe sizes for the original fixtures may not be adequate for current usage. You may have to change the old pipes and fit adapters between the new larger pipes, and the faucets and waste lines.

Many warehouses with plumbing parts stock old enameled-iron clawfoot bathtubs, marble vanity tops, tank toilets, and decorative painted porcelain washbasins. Good reproduction brass and porcelain fittings are now available, so that it is possible to produce a traditional bathroom without sacrificing function or convenience. Some hybrid concessions to modern showers, thermostatic valves, grab bars, and other such amenities are unavoidable.

Gas Piping

In the second half of the nineteenth century, gas was the most common form of illumination in urban areas. When electricity was introduced, many buildings were equipped with dual systems. Gradually, as gas illumination was phased out and supplanted by electricity, the old piping systems were abandoned. Often the new wiring was pulled through the old gas piping. Some gas service was retained for domestic hot-water heaters and for cooking stoves. At the turn of the century, coal was the most popular heating fuel, but some gas-fired fireplace ceramic units, or logs, continued to be used. Gas-fired heating systems again became popular after World War II.

For safety, as well as for economic reasons, it is essential that old gas piping systems be thoroughly checked before any interior demolition occurs. Occasionally, one finds stubbed-off nineteenth-century wall sconces that have been improperly disconnected from a gas service that is still in use for domestic hot water and cooking. Most restoration projects use special "flickering" electric light bulbs to simulate gaslight or fireplace gas logs because of potential fire hazards.

Fire-Suppression Systems

Most historic structures are not fireproof, and therefore automatic fire-suppression systems may be required. The essential purpose of these fire-suppression systems is to provide occupants with an opportunity to evacuate a burning building. If the historic structure itself or its contents are to be salvaged, a far more complex alarm and fire-protection system must be devised. Major problems in restoration projects are encountered when installing and concealing the necessary plumbing and sensing-and-control wiring, all of which disrupt historic finishes. (Alarm systems are covered in depth in Chapter 9.)

Sprinklers. If an automatic sprinkler system is introduced into an old building it may be necessary to increase the size of the water supply piping to provide the required pressure. Old industrial sprinkler systems may have to be adapted or replaced when former lofts or manufacturing buildings are converted to residential or commercial uses. Over the years, many old sprinkler systems have been poorly maintained and may be inoperative because of rusted or painted heads. It may also be necessary to rezone or redistribute the location of the sprinkler heads to conform to new layouts. False alarms and freezing can cause extensive water damage, so that dry systems or new chemical fire-suppression systems may be more appropriate.

Halon. Halon gas has been developed as an alternative to fire-suppression systems using water or chemical foam. Halon smothers fire and leaves no residue. Because it is a gas, it must be maintained under pressure and the storage tanks must be recharged periodically. In order to maintain the proper halon concentration, the areas it can service are limited in size and must be compartmentalized. These limiting conditions do not always suit the layout and character of existing historic interiors. Although halon is quite expensive, its installation requires much less plumbing and results in less disruption of existing interior finishes than sprinkler systems do.

Because of its smothering effect, all persons must be evacuated before the gas is released, and ventilating equipment is required to provide a purging cycle following an alarm. Halon systems are now common in storage areas that contain critically important or irreplaceable equipment, records, or objects (such as computers, telephone-exchange systems, and works of art). Halon is ideal for attics within wood gabled-roof framing, such as those on most old churches, even those with masonry walls. The halon system in an attic may be combined with conventional water sprinkler systems in a sanctuary and other assembly spaces.

MECHANICAL SYSTEMS

Old buildings have to be brought up to the twentieth-century standards for comfort, health, and safety. Many of these standards are required by law, some are by choice. Mechanical systems, such as piping, ducting, and wiring for heating and air conditioning, have to be fitted into an existing structure with the least damage to the aesthetics of the interior, a task calling for sensitive design, skilled craftsmanship, and an awareness of potential side effects.

Mechanical systems are not expected to last more than a fraction of the life of a building. Make as few changes on the building's fabric as possible because it will probably be redone in another 30 or 40 years.

Despite the attention to authenticity and historic integrity, very few preservationists would suggest doing a restoration lacking central heating, running water, and electricity. Thus, we are immediately faced with the contradiction of a restoration that is historically "accurate" while it deftly conceals modern illumination and systems for climatization and security. In museum settings a certain amount of inconvenience may be accepted for the sake of appearances, but in the case of an adaptation to a contemporary use this cannot be tolerated. Obviously, the earliest buildings of the colonial period had very primitive conveniences. Toward the mid-nineteenth century the precursors of modern mechanical systems began to evolve. The indoor bathroom replaced the outhouse, the gravity air heating system replaced the fireplace and the stove, and illuminating gas replaced candles and oil lamps. Many old buildings have acquired these new mechanical systems as they were developed. It now becomes a matter of selection as to which of these layers of technology should be stripped away and which should be preserved or restored. It is possible to return a building to its eighteenth-century appearance by replacing nineteenth-century cast-iron stoves on hearths with restored, original fireplaces, and also to provide modern forced air heating.

Installing and concealing modern systems can be very difficult depending on the type of construction and the architectural character of the particular building. Finding nooks and crannies in which to hide new mechanical elements is not always easy, especially in buildings with structural framing systems exposed on the interior (see Figures 7-52 and 7-53).

Ducts or pipes can usually be concealed in wall cavities or under floors, but when they cannot be hidden they have to be located in the least obvious places and painted to visually merge with the background (see Figure 7-54). Heating and cooling units that cannot be concealed should be selected for the quality of their design and not because they have Queen Anne legs. The location of these units demands sensitive placement.

Before deciding to put a new heating system into a building or adding an air conditioning system, be sure to calculate the effects of the new artificial climate you are considering. This is particularly important with old buildings that have not previously had their humidity greatly changed by a mechanical system.

If the interior climate becomes too cool in summer, the humid air will condense and penetrate any walls not adequately protected with vapor barriers. Similarly, in winter, moisture from humidifiers will penetrate walls and rot the wood or peel the paint and wallpaper. Also, if the humidity is drastically reduced, the building's framework can shrink, causing plaster surfaces to crack and window and door frames to separate.

Figure 7-52 City Hall, Oswego, New York. New mechanical systems are being installed in exposed framing and wall areas prior to the restoration of interior finishes.

Figure 7-53 City Hall, Oswego, New York. A new plaster ceiling, reproducing original molding profiles, conceals the installation of modern mechanical systems.

Figure 7-54 City Hall, Oswego, New York. Close-up of the restored plaster ceiling reveals a linear air diffuser slot concealed along the edge of the new decorative plaster molding. This is a far more attractive method than the more conventional practice of installing metal grilles and diffusers.

HEATING SYSTEMS

In approximately 300 years of American building there has been a considerable evolution from the fireplace to year-round controlled environments. An awareness of the gradual stages of development will help in guiding appropriate preservation efforts.

Heating systems, except electric ones, require a flue to remove burned gasses. The flue of original chimney will probably be adequate for a new furnace, fireplace, or heating stove if it complies with current building codes. The exterior of a masonry chimney can be repointed and the flashing renewed if necessary. Make sure that the thickness of the chimney's walls are up to code standards, especially if you install a furnace with higher flue gas temperatures than the existing burner.

The interior of the flue pipe is hard to examine. Experienced chimney sweeps can feel if there are breaks in the flue lining, but it would be prudent to reline the chimney with a stainless steel pipe or fireclay sections. While checking the chimney, see if the structural timbers close to it are charred. If they are, cut back the framing members further from the chimney.

Fireplaces

The earliest and most basic heating device is the fireplace. Many old fireplaces are unsuitable by modern safety standards because they do not have lined flues, adjustable dampers, or firebrick-lined chambers. Until the availability of highly fired vitreous clay or refractory brick, which was developed for the nineteenth-century iron foundries, the interiors of fireplaces were coated with a cement rendering that insulated the brick against excessive heat. It was far easier to replace a cracked cement coating than to rebuild the chimney.

Although builders in the last century understood how to design fireplaces to make the fire draw well, there are some recent improvements to dampers and throats that should be considered. Another improvement is to put a stainless steel, 1/2-inch mesh over the top of the chimney to arrest sparks and prevent birds from falling in.

Cast-Iron Fireplace Backs

The first settlers brought with them an early attempt to increase heating efficiency. A cast-iron fireplace back was mounted against the rear wall of the firebox, where it would absorb a portion of the heat and then radiate it back into the room. Adapting this principle, Benjamin Franklin reasoned that if the entire fireplace were lined with iron sheets it would radiate that much more heat. Using this notion he experimented with moving the iron firebox out onto the hearth and created the Franklin stove.

Iron Stoves

The stove proved to be far more efficient for heating than the fireplace, and became immensely popular. A kettle of water kept on the stove provided an elementary humidification device as well as a quick source of tea. As iron stove designs evolved, it was discovered that even the stovepipe could yield additional heat and so it was run, exposed, from the stove into the former fireplace throat above the height of the mantle or directly into a chimney flue. A circular hole about 7 inches in diameter located about 6 feet above the floor and covered with a metal "pie plate" is evidence of a former stove location.

In rural buildings, which might only have had a parlor stove in addition to a kitchen range, transfer grilles were placed in walls and ceilings to extend the heat to the remaining rooms. Reproductions of these old transfer grilles are now being manufactured and can be used as diffusers directly attached to modern ducted forced air heating and cooling systems.

Coal Stoves

Perhaps the most significant factor in the evolution of nineteenth-century heating methods was the shift from wood to coal as the principal fuel. Not only was coal more efficient and produced less ash, but it was also easier to transport. The coal grate was more compact and required a smaller flue than the woodburning fireplace. Purchasers of old houses mistakenly assume that all fireplaces are woodburning, and only when smoke fills a room do they discover that the smaller flue was for a coal grate and that it does not draw well for wood fires.

CENTRAL HEATING

Until late in the nineteenth century when the concept of central heating became more practical, most buildings were still heated on a room-by-room basis requiring independent flues and chimneys for each room served.

Gravity Air Systems

One of the first methods of central heating was the gravity air system, which depended on the natural convection effects of hot air and cool air. A wood or coal furnace in the cellar would heat the air in a plenum chamber connected by sheet metal tubes or ducts to risers set in chases in the walls. Both floor- and wall-mounted decorative bronze grilles concealed a set of adjustable louver vanes, which could be regulated to control the desired flow of hot air. Many of these old coal-fired gravity systems were considered dirty because coal dust infiltrated the plenum and was distributed along with the warm air. If old gravity ducts are to be refurbished for a forced air system, it is wise to first thoroughly vacuum the ducts. The distribution of the warm air was often uneven and could be drafty.

Because of the need for ducted plenum space, the grille locations are set in the solid interior walls and not directly in front of window openings where there is the greatest heat loss. As a result, many of the gravity air systems were abandoned in favor of coal-fired steam radiation systems. Whereas gravity systems were almost silent, steam systems are constantly hissing, clanking, and banging.

Old buildings with extant gravity systems can be converted to modern forced air combined cooling and heating systems. Because the modern ducted air systems contain fans and are maintained under pressure, they require a smaller cross-section than the original gravity systems. This differential can provide a convenient concealed space for installing modern piping and wiring. Replacement grilles with booster fans can augment an existing gravity system and provide superior heat distribution.

Steam Systems

Coal-fired steam heating systems came into use at the end of the nineteenth century. The early systems were not automatically stoked, required periodic ash removal, and needed the addition of boiler water. Thermostatic control was nonexistent, so that overheating was a problem. As technology developed, fuel sources changed first to oil and then to natural gas.

Old steam systems can be restored if there has not been too much rust or corrosion in the piping and if the radiators are not cracked (Figure 7-55). It is not always possible to repair damaged cast iron, and finding duplicates of vintage designs ranges from frustrating to futile. New cast-iron steel radiators are available, but their profiles are generally slimmer and lower than the old ones. Depending on the quality of the water supply, steam can be extremely corrosive. It creates chlorides that attack the cast-iron and steel elements of the system.

Modern furnaces are more compact and more efficient in their use of fuel, but some of the inherent disadvantages of steam—dryness and lack of variability in temperature—are not easily or inexpensively overcome. For buildings with seasonal or occasional usage there is another problem. The system cannot be temporarily shut down, unless it is fully drained, without risking freezing, broken pipes, and flooding. In some cases, a combined system may be the most satisfactory choice. Moreover, in some regions of the United States it is increasingly difficult to find skilled plumbers and the ncessary replacement parts to make continued maintenance practical. The radiators themselves may be retained for the sake of period authenticity, but a completely modern system may have to be provided to supply the actual climatic needs.

Figure 7-55 Concealing new mechanical systems, restored bank, Canandaigua, New York. By creating a pattern of unobtrusive slots, modern radiators have been introduced. But the traditional character of this nineteenth-century interior with its scored wood "novelty" board wainscot has been maintained.

CHAPTER EIGHT

RESTORING INTERIORS

FLOORS, WALLS, CEILINGS, FURNISHINGS

Defining what is American about interior design and the decorative arts is as elusive for preservationists as it is for museum curators. Just as the American people are an amalgam of nationalities, eclectic borrowing from foreign cultural influences has been a pattern from the earliest colonial period.

New England traders were importing not only tea and spices from China, but were also bringing back porcelains, furniture, and elegant hand-painted wallcoverings. Such exotic influences were sources of architectural inspiration. Thomas Jefferson's design for the University of Virginia at Charlottesville combined wood balustrades with Chinese Chippendale fretwork with classically inspired colonnades.

Because of the status that imported luxuries imparted to the wealthy, local craftsman fashioned naive copies of imported luxury goods for the less affluent. Exotic woods and fancy marble were mimicked in paint finishes by skilled grainers. Humble pine cabinets were rendered as ebony or mahogany. Plain slate fireplace mantles were transformed into richly veined "marble"—the Formica of the eighteenth century.

Taste and fashion are cyclical; popular styles become passé and then are rediscovered. In recent years a vogue for the plain "country look" has led many antique dealers to strip fine eighteenth- and nineteenth-century painted furniture to reveal the "mellow" wood beneath the faded finishes, thus diminishing their real value. Similarly, the novelty of exposed brick walls has caused many overzealous owners to destroy historically intact interiors and substitute so-called restored interiors. Massive golden oak bars set against exposed brick walls, with potted ferns hanging in the windows, may have great nostalgic appeal but, in fact, the ensemble bears no relation to nineteenth-century saloons.

Even when original interiors are left intact, many architects and designers are content to paint out all traditional detailing, moldings, walls, and ceilings and create a uniform white envelope that is intended to be a striking contrast for the modern furnishings in the room. And, though it may not be practical or desirable to totally re-create a historic interior, there is a growing sentiment that the original architectural interior "shell" should retain the integrity of its materials, colors, detailing, and texture.

EARLY STRATEGIC CHOICES

The interior of an old building usually has been subjected to more changes than the exterior. Quite often there is little extant physical evidence to serve as a guide to restoration. In this circumstance the architect has several acceptable alternatives.

1. **Create a modern interior.** From a practical standpoint the loss of the original interior finishes can be advantageous, permitting greater latitude in the choice of a new use. Greater flexibility in the choice of colors, materials, and layout is another advantage. The new interior may be developed so that the old exterior retains its original appearance.

2. **Create a conjectural re-creation of the original interior.** Many old buildings are similar to surviving prototypes, so that a new interior can be fashioned to approximate an old one, or missing portions of it can be restored, without the need for exhaustive research. This approach requires considerable restraint, since it is tempting to opt for an over-elegant version of the past.

3. **Create a pastiche of old and new.** In many cases, parts of the original decorative elements and finishes remains intact and can be incorporated into a new design. Fancy cast-iron columns, pressed tin ceilings, stained glass windows, wood paneling, and mosaic tile floors are typical nineteenth-century features. These elements can be inventively combined with modern forms and finishes with no pretense of restoration. It is also possible to replace missing architectural detailing with old decorative elements by searching in parts warehouses.

Quite often either budgetary or programmatic considerations dictate the choices between these alternatives in an actual situation. In the case of properties listed on the National Register of Historic Places, alterations must adhere to the Secretary of the Interior's standards. These standards are quite explicit regarding the retention of significant original interior features and finishes and serve as mandatory guidelines for those seeking certified rehabilitation status in order to qualify for Internal Revenue Service tax incentives.

Restoring the Original Look

Researching the probable finishes and furnishings of a house can be very frustrating, since records of furniture layouts were rarely kept. When detailed inventories or bills and receipts for purchases are available, they are, at best, annotated lists and not plans or drawings that indicate layout or arrangement. Therefore, in the absence of archival documentation, you must rely on local tradition as a guide to restoring the interior and furnishings. In most areas of the country, local museums, historic societies, and houses are repositories of invaluable information about regional preferences.

Regional traditions have strongly influenced architectural and decorative tastes. From Philadelphia south, the focal point of a fashionable drawing room was the fireplace mantel with a richly carved overmantel that typically incorporated a central panel containing an oil painting or a mirror. By contrast, the architectural finishes of New York houses, until the late nineteenth century, included only the fireplace mantel, with owners selecting mirrors, paintings, et cetera to hang above it.

Old buildings are changed subtly over time to suit both the needs and tastes of their owners and occupants. One must therefore assume that even in the unusual circumstance of continuous occupancy by the same family, some furnishings, decorative objects, and color schemes were acquired, arranged and rearranged, and even disposed of. Even the most "authentic" restoration can never re-create a room exactly as its original occupants experienced it.

Because exact color descriptions rarely exist, a paint analysis must be used to verify the range of possibilities that may lie beneath accumulated layers of paint (see under "Paint" in Chapter 7). Quite often these investigations result in some discoveries that shock our modern sensibilities. A recent exhaustive paint analysis at George Washington's celebrated Mount Vernon revealed a palette of bright, clear blues and greens that was far from subtle. The curators at Mount Vernon have had the courage to re-create the entire interior, including the upholstery to match.

In nonmuseum restorations, an individual owner may have strong likes and dislikes and refuse to go along with the original color scheme. In these circumstances, selecting from a palette of other appropriate colors seems a valid alternative. We must assume that people in the past had a choice. There is far less threat to historic integrity in varying the color scheme than in ripping out moldings and mantels to suite contemporary whims.

Protecting Interiors before Construction Begins

Once the research has been completed and you have a restoration plan, and before any construction takes place, a complete protection plan must be organized to prevent damage and theft. Everything that is easily removable, such as decorative hardware, light fixtures, metal grilles, and fireplace grates, should be labeled and placed in safe storage. It is also a good practice to photograph everything in place before dismantling for storage.

Fragile decorative features, such as plaster ceiling medallions that cannot be removed, should be protected in place prior to any demolition. An additional but costly precaution is to take molds of special plaster decorative elements so that replacement casts can be produced. Modern casting techniques, such as those for fiberglass, are often superior to the original plaster since they are lightweight and moisture resistant (see under "Plaster" in Chapter 7). During the course of protecting what there is, it may become apparent that certain architectural elements are missing. Some others may require replacement because of proposed alterations. Often there is no practical alternative to destroying portions of the original interior finishes in order to install modern wiring, plumbing, and mechanical systems.

Locating Missing Pieces

Because of the tremendous interest in historic preservation, a new antique trade has developed in salvaged architectural elements. Demolition crews once totally destroyed old buildings; now there is more careful dismantling and removal of decorative elements. Ironically, many buildings today are worth more in pieces than intact. Several early Frank Lloyd Wright houses have been salvaged and parceled out to museums and collectors, none of whom would have been willing to preserve the buildings in situ.

There is, however, a legitimate need for parts depots to supply missing elements for restoration projects. Nonprofit groups organized the Parts Warehouse in Albany, New York, and the Baltimore Salvage Depot in Maryland. Several other cities have established parts warehouses to assist owners who are restoring historic properties. These noncommercial operations are akin to legitimate adoption agencies; they are careful about their sources and scrupulous in their placements.

Before proceeding with more specific aspects of the restoration and recycling of historic interiors, preservationists should be familiar with the typical construction methods common to American buildings. An overview of traditional methods of floor, wall, and ceiling construction follows. Elements of interior design are discussed at the end of this chapter.

TRADITIONAL WOOD FLOORS

Most traditional floor construction in the United States is wood-framed. In the early colonial period, large trees were abundant and sawmills scarce. Wide planks were commonly available and were used for flooring. Today wide planks are rare and highly prized.

Problems with Old Wood Floors

Often in the early colonial period the planks were cut thick and spanned the heavy timber framing with no subflooring. Over time there has been a tendency for these broad planks to shrink, causing gaps to appear. Since the early planking was neither shiplap nor tongue-and-groove, shrinkage also caused the planks to warp, or cup. This was not very satisfactory, especially when an occupant saw light, heard noise, and found dirt coming through the ceiling from the floor above.

Repairing Wood Floors

One method of closing the gaps is to insert wood filler strips, but it is difficult to blend this repair with the mellow, aged patina of the original floorboards. A less satisfactory approach is to cover the floor with another layer of wood or some other flooring material, which conceals the wide planks. It is also possible to lift the planks, lay a new subfloor directly on the beams, and then reinstall the broad planks. All of these remedial methods are complicated by the fact that many old buildings have settled unequally and have floors that are warped and off-level.

Some portions of an old wide plank floor may be so worn or damaged that patching or repair may be required. In this case it may be necessary to remove a section from a closet, secondary space, or attic to use for patching. Ordinary wood strip flooring or plywood can be used as a replacement for the removed portions.

Providing for Expansion. All wood is subject to swelling and shrinkage due to seasonal variations in humidity. Some allowance for this movement must be made at the edges of the room, or the expansion of the individual planks will cause buckling and unevenness of the flooring surface. In traditional construction this allowance or gap is concealed by the baseboard or shoe molding that runs around the room.

Refinishing Wood Floors

Patching with different woods can cause difficulties in achieving uniform color and finish. Although wax was the traditional method of floor protection and finish, today surface coatings of polyurethane offer easier maintenance and give soft woods, such as pine, greater resistance to scuffing and marking as well as to staining from moisture.

To successfully apply a coating of a new finish, the old finish must be removed. This requires careful sanding. Machine sanding may ruin the subtle unevenness of an old wood floor and create additional problems. All gaps and nail holes must be patched and skillfully stained prior to applying the polyurethane coating.

In the usual sequence, floor finishing is one of the last jobs after the carpentry, plastering, and painting are done. Even though vacuum attachments are used on the floor sanding machines, a fine layer of dust usually remains after sanding. And invariably some touching up of painted baseboards and trim is necessary after the floor finisher leaves. Thus, it is better to install fine wallpaper, fabrics, and elaborate light fixtures after the floor has been sanded and coated.

Wood Floor Finishes. Most common finishes for wood flooring—wax, mineral oil, varnish, and shellac—are clear or stained so that the wood grain can be appreciated. However, paint is sometimes used. On outdoor porches and steps and other heavy-use areas, paint is often used. As part of the recent fashion for the "country look," a revival of traditional spattered-paint or stenciled designs is immensely popular. However, stenciled floors are extremely fragile and vulnerable to wear. Applications of low-luster sealers or polyurethane help in prolonging the serviceability of these decorative but impractical floors.

Combine Wood with Other Materials

In very elegant formal installations, such as the Frick Gallery in New York City where a wood floor is set in a marble surround, a strip of cork, which compresses easily, is inserted around the edges of the wood flooring to serve as an expansion joint. One of the difficulties of abutting dissimilar materials, such as marble and wood, is that mopping with water, which is the most effective cleaning method for marble, causes staining and swelling on wood flooring.

Wood Strip Flooring

From the nineteenth century to the present the most common flooring has been narrow strips of pine, cedar, maple, or oak. Typically, strip flooring was either shiplap or tongue-and-groove over a wood subfloor of rough-sawn planks running perpendicularly or diagonally in relation to the framing joists beneath.

The individual flooring strips are toenailed into the subfloor to conceal the nailheads. This makes it difficult to remove or replace a damaged strip in the middle of the floor without disturbing the surrounding strips. Sometimes patching materials can be found in closets or concealed beneath trim.

MARQUETRY AND PARQUET

As nineteenth-century tastes shifted from the austere neoclassic to the more decorative baroque French and Italianate modes, wood strip flooring was considered too plain. In the principal entertaining rooms, corridors, and stairhalls, ornamental borders of inlaid exotic woods in geometric patterns or garlands framed a central panel of wood strip flooring. Those who could not afford genuine hand-cut marquetry had to settle for stenciled imitations using varicolored stains. Often the central field of the floor within the decorative border was made of oak strip flooring because it was to be covered with carpet.

Figure 8-1 Marquetry, Villard Houses, New York, New York. At the end of the nineteenth century the fashion for wall-to-wall Wilton carpet reverted to the more traditional preference for fine old rugs, along with other antique furniture, tapestries and objects d'art collected on the grand tour abroad. Intricate marquetry borders were inlaid with exotic woods of contrasting shades, framing a central panel of parquet over which the rug was placed.

But in some cases elaborate parquet patterns were worked into the central panel. Today we think of parquet as an alternating checkerboard of square wood tiles. The original "parquet de Versailles," which nineteenth-century tastemakers copied, was far more elaborate, consisting of herringbone patterns and geometric designs of interlocking individual wood blocks. Expensive hardwoods and fruitwoods were combined to produce rich contrasts of color and graining.

Repairing and restoring inlaid marquetry borders and parquet floors is difficult because of the many small pieces involved (Figure 8-1). Over the years a combination of normal foot traffic and furniture movement cause individual pieces, or inlays, to become loose or dislodged. Conventional floor sanding machines are not recommended for these types of floors. The patterns in stained imitation marquetry may be totally erased. It is essential that only small hand-held sanding machines be used, and only after all the loose pieces are secured and missing elements replaced. Once the repairs are made a protective coating of polyurethane may be applied.

TRADITIONAL FLOOR COVERINGS AND CARPETS

In most old buildings, floor surfaces may be in poorer condition than the other interior finishes because of prolonged usage and general wear and tear. In seeking to extend the practical life of a structure or to adapt it to new uses, it is helpful to become familiar with the historic development of the traditional floor-covering materials.

Carpets and Rugs

From the earliest colonial period, carpets were acquired through trade in the East. Turkish and Oriental carpets were often so highly prized that they were draped over tables and not walked on. In order to protect those carpets used on floors from wear and stain, floor cloths made of oilcloth cunningly stenciled to imitate marble or trompe l'oeil effects were overlaid. (See under "Modern Floor Coverings" later in this chapter.)

While the well-to-do collected Oriental carpets, citizens of more modest means contented themselves with braided or woven rag rugs made of old garment scraps. In the early nineteenth century, the invention of the Wilton loom in England revolutionized carpetmaking, transforming a slow and tedious handicraft into a modern, mechanized process. For the first time, wall-to-wall carpeting was a practical decorating method. The narrow 27-inch patterned strips and borders could be sewn together to fit any room size, corridor, or stair. The Wilton process made possible an infinite variety of patterns, from intricate geometric motifs to huge swirling designs of flowers and foliage.

During the so-called Brown Decades, from the 1840s to the 1880s, the fashion for wall-to-wall carpeting made elaborate wood flooring unnecessary. Rooms with baroque carved plaster ceilings and bold wood detailing and trim have surprisingly plain pine flooring. These areas were obviously meant to be concealed by Wilton carpets.

Hot summers in the generations before air conditioning meant leaving the windows open. The gritty atmosphere produced by coal-fired nineteenth-century industries left its traces on carpets everywhere and effective cleaning methods were very limited at this time. Because the wall-to-wall Wilton carpets were pieced together, stretched, and then tacked in place, it was impractical to remove them seasonally. Straw rugs, or matting, were placed over carpeting for the summer as a way to minimize soiling the carpets. The richly upholstered damask and velvet furniture was covered with cotton slipcovers and heavy draperies were taken down and replaced with lighter summer substitutes.

In monumental and public buildings where marble staircases were de rigueur, runners were fitted down the center of the steps to make the marble less slippery and clattery. A system of brass rods held in place by brass anchors restrained the carpet at the intersection of each riser and tread. This method is clumsy but effective. The use of Velcro backing strips also prevents carpet movement with no damage to the floors.

CERAMIC AND MOSAIC TILE

During the colonial period, ceramic tiles were used in the hearth area as a fire-resistant material. The Spanish settlers used blue-and-white decorated azujuelos and the Dutch used imported delft tiles. The traditional use of ceramic tiles surrounding the fireplace opening was supplanted by slate in Classic Revival interiors of the 1830s and by marble during the Italianate vogue around midcentury.

The advent of indoor plumbing in the 1840s provided a new demand for ceramic tile, first as floor tile and soon after as wall tile. Small mosaic tiles in octagonal or hexagonal shapes could be easily fitted in nooks and crannies with a minimum of cutting. Later, elaborately modeled, painted, and glazed tiles and accessories were available to decorate sumptuous American bathrooms. Whereas in England, traditional potteries, such as Minton and Royal Doulton, manufactured everything from hand-painted soup tureens to commode sets, American companies tended to specialize in hard porcelain products: tile, plumbing fixtures, and fine china tableware services.

During the 1860s and the remaining Brown Decades, imported Minton tiles in dull earthen hues were widely installed in hearth and fireplace surrounds as well as in areas subject to dampness and high traffic, such as entry vestibules and corridors in public buildings, churches, and railroad terminals (Figure 8-2). Minton tiles were decorated with arabesques, plant forms, and Gothic and heraldic motifs that could be composed in geometric patterns matched with borders. During the nineteenth century, an American ceramic tile and brick industry developed in Ohio and other regions with substantial clay deposits. The domestic industry soon copied the imports and made them widely available at less cost. Minton has now begun to reproduce many of its original nineteenth-century patterns, but so far the U.S. tile industry has not followed suit.

Figure 8-2 Decorative tile fireplace, Ballantine House, Newark Museum, New Jersey. In the colonial period, when the fireplace was the sole source of heat, delft or decorative tile was limited to the surround framed by the wood mantel. By the turn of the century the decorative tiles lined the entire firebox which contained a "gas log" to provide a cozy warmth to supplement the central heating system.

During the latter half of the nineteenth century, small-scale, solid-colored ceramic mosaic tile floors were popular in commercial buildings, drugstores, ice cream parlors, and saloons, where they could be maintained easily with a mop and a bucket of water. Identical tiles are still available, with the added convenience that they are now preset on sheets that facilitate installation.

At the turn of the century, as many wealthy Americans made the grand tour of Europe, they returned with a desire to embellish their homes and communities with the rich Venetian mosaic and inlaid marble ornamentation they had seen abroad. They were eagerly assisted in this campaign by the first generation of American architects to have been trained in Paris at the Ecole des Beaux-Arts. Monumental structures built in the Richardsonian Romanesque, Gothic Revival, and Beaux-Arts styles provided ample opportunity for immigrant Italian artisans to apply their traditional skills. Churches, public buildings, private clubs, and mansions were decorated with mosaic floors, vaults, domes, fireplaces, and fountains. Inspired by the popular European Art Nouveau movement, American designer-craftsman Louis Comfort Tiffany produced dazzling effects by combining traditional Venetian glass mosaics with iridescent favrile of his own invention.

Traditional mosaic installations have endured for centuries and require only periodic cleaning to restore their luster. Sometimes cracks appear because of water penetration, vibrations, or structural movement in the surfaces to which the mosaics have been applied. A skilled artisan can remove the affected portions of the mosaic and reset them after the structural repair is made.

In the 1870s and 1880s, the English Arts and Crafts Movement, championed by William Morris and John Ruskin, and the Art Nouveau style on the Continent inspired a number of artists in the United States to experiment with crafts, particularly ceramics and tile making. The Rookwood and Grueby potteries in Ohio, and Moravian Tile Works in Pennsylvania, were pioneers in creating glazed ceramic art tiles. In contrast to the classical imagery of the Beaux-Arts style, naturalistic forms and folklore predominated. Once commonly available through catalogs, the art tiles of these talented turn-of-the-century American artisans are now highly prized collectors items and museum pieces, and many are preserved as part of landmark structures and interiors.

Unglazed terra cotta waxed to a leathery patina became popular as a part of the Arts and Crafts, Mission, and Bungalow styles. Reproductions of primitively painted and glazed Mexican, Spanish, and Italian tile were lavishly used by Addison Mizner in his fantasy villas in Palm Beach and Boca Raton, Florida.

Most of the varieties of tile described here are still being produced today or have acceptable facsimiles. Some parts warehouses sell salvaged decorative tiles, and even if an exact match is not available, something can be found as a substitute.

MODERN FLOOR COVERINGS

One of the great innovations of the eighteenth century was the floor cloth made of an oilcloth, usually canvas, like a fisherman's slicker. It was a lightweight, inexpensive, water- and-stain-resistant covering that could be laid on top of a precious Oriental rug. Placed under a dining room table, it would protect the rug not only from soiling but also from the damage caused by frequently moved chair legs. The humble purpose of the floor cloth was disguised by stenciled or painted decorations such as those that imitated checkerboard-patterned marble floors.

Entry halls and even porches were somctimes covered wall-to-wall with canvas and then painted to imitate marble. Thus, the floor cloth is the precursor of modern linoleum and vinyl seamless flooring.

Inlaid Linoleum

Hand-painted floor cloths were still relatively fragile and subject to wear. In the nineteenth century, linoleum, a durable and washable material, was made in large sheets by pressing a mixture of heated linseed oil, rosin, powdered cork, and pigments onto a burlap or canvas backing. Early linoleum was monochromatic or, at best, marbleized. Linoleum of contrasting colors was inlaid in patterns and as borders. Soon, cheaper imitations appeared with printed designs on thinner material that was available in long rolls.

Linoleum was commonly used in high-traffic areas, such as corridors and courtrooms. It was often applied over earlier wood floors in an effort to increase serviceability and reduce maintenance. In most places, linoleum was pasted directly on old floor surfaces, including tile and marble, making its removal a tedious process. The popularity of linoleum continued until the 1950s, when the sheet vinyls supplanted it. Most linoleum available today is manufactured in Europe.

Vinyl Flooring

Sheet vinyl is a petroleum-based product that is part of the plastics revolution of the post-World War II period. It is much more resilient than linoleum and does not require a backing. Vinyl is a molded product capable of reproducing fine detail and accurate coloring and is, therefore, a useful product in restoration.

ARCHITECTURAL WALL FINISHES

It is not always easy to distinguish between architectural finishes that are an integral part of a building's fabric and applied interior finishes. Tinted plaster with a sand finish is an architectural finish; wallpaper is an applied finish. However, stenciling "applied" to walls is nonremovable—unlike wallpaper—and is, therefore, classed as an architectural wall finish. These distinctions may seem academic, but they are useful as organizational labels.

Stone and Marble Masonry Finishes

In institutional and public buildings, stone and marble have been used as paving because of their durability and resistance to soiling. To minimize damage to adjacent walls, a base or wainscot of the same stone or marble is often provided. In the early colonial period the choice was usually limited to what the local quarries offered, with the coastal areas generally better served. Importing marble was costly and slow, and as a result, marble achieved luxury status. Those who could not afford the real thing would seek out imitations. Even in the nineteenth century, when marble and stone became more abundant, inexpensive facsimiles were still very popular. Slate was grained to give the appearance of exotic marble; cheaper still, but not as serviceable, was grained, painted wood.

In recent years state preservation offices and local historic groups have researched and documented the original regional quarries and sources of stone. Although many of these quarries no longer function, a surprising number still do, and others nearby produce material similar enough to use as patches or replacements.

Cleaning Interior Marble and Stone. The traditional stone and marble cleaning method—applications of Keene's cement and poultices of various compositions—is now joined by newer processes. Because interior surfaces are not exposed to the weather, the identical marble or stone used on the exterior will present very different cleaning and maintenance problems. Some of the stone cleaning methods that have been developed for exterior masonry are impractical for interior applications because they are too messy and destructive of other finishes. Because of the experimental nature of some of these newer methods, it is wise to test a sample in an inconspicuous location before undertaking a large project.

Ceramic Tile, Terra Cotta, and Brick

Most of Europe had exhausted its timberlands by the eighteenth century, and urban construction was predominantly masonry. The early colonists adapted familiar traditions to the indigenous materials and climatic conditions of the New World. Masonry in any form takes more "manufacturing" than architectural wood elements, and because of that, it was imported at first. But masonry remained expensive even when it was locally produced. Dutch blue-and-white delft tile was prized as a decorative, fireplace surround. In the Spanish territories blue-and-white *azujuelos* decorated kitchens, fountains, and patios.

Many regions of the United States contained clay deposits and in the nineteenth century these resources were extensively developed and exploited. The Victorian taste for decoration and the advances in indoor plumbing gradually led to the status "tiled" bathroom.

At the turn of the century, the American terra cotta industry produced amazing imitations of custom stone carving and finishes in an extraordinary range of rich glaze colors. Most of the interior installations in lobbies, for instance, have come through the ravages of time in perfect condition, unlike many of the deteriorated exteriors of the same buildings.

The 1920s revival of Mediterranean styles in Palm Beach and Hollywood popularized the vogue for decorative tiles. The flat geometric aesthetic of Art Deco in the 1930s found frequent expression in glazed terra cotta tile that adorned every kind of structure from gasoline stations to movie palaces (see Figure 8-3). Glazed and molded brick also celebrated the "streamlined" World of Tomorrow look of the 1939–1940 world's fair.

Figure 8-3 Glazed terra cotta, I. Magnin Store, Oakland, California. Brightly colored, nonfading glazed and molded decorative terra cotta offered an attractive and distinctive exterior treatment for the fashionable merchant, at less cost than stone or marble. In some climate zones these facades have not fared well over the years and now require extensive repair and restoration.

Because of the permanence of traditional tile setting methods it is almost impossible to remove tiles without breaking them. Although imported delft, Minton, Mexican, Italian, French, Spanish, Portuguese, Chinese, and Japanese tile abound, finding replacements for some old tile is a challenging but by no means futile pursuit.

Stucco, Textured Plaster, and Molded Plaster

During the late seventeenth and the early eighteenth centuries, only the wealthiest Dutch patroons and English Southern planters could afford to indulge in decorative plaster. Using decorative or molded plaster was limited to ceilings in principal rooms and was rarely applied to the walls.

Certainly in the early colonial period there was no desire to create a picturesque, rustic, or antique effect. Any crudeness or primitiveness of finish was a result of the technical limitations of local craftsmen. In rural areas, horsehair and other animal fibers were used as binders to reinforce the soft plaster finish. Because these early plaster walls were so fragile, it became common to install a molded or flat wood chair rail to protect the finish from the occasional scraping of movable furniture.

The interior plaster walls of most ordinary buildings were given a periodic coating of lime-based whitewash or, in some rural areas, milk-based paints. Stenciling and painting directly on the whitewashed walls was an inexpensive, popular means of achieving a decorative effect. The wealthy, particularly those involved in the China trade, imported charming hand-painted wallpapers with birds, trees, and flowers. Silk-screened and hand-blocked scenic and patterned wallpapers and borders from France and England were also available.

Most of these wall coverings were applied with wheatpaste directly to the plaster. It is very difficult to remove these early wallpapers intact in order to restore the crumbling old plaster walls behind them.

The severe Federal and Classic Revival styles eschewed decorative plaster on wall surfaces and restricted carved architectural detail to wood elements of door and window frames and fireplace mantels. The preference for the historic revival styles, such as the Italianate and Gothic, began a trend to more elaborate decoration, which eventually left few surfaces untouched. Along with the romantic aesthetic of the period, there was a deliberate attempt to recreate the ambiance of far-away and ancient cultures. Plaster and paint were ideal and convenient media for economically simulating instant antiquity and the richly ornamented treasures of the past.

Travel abroad by wealthy Americans, as well as several major international expositions at the close of the nineteenth century, increased the desire to "bring home" exotic and picturesque places. Aided by the popular Arts and Crafts Movement, the stage was set for the deliberate simulation of period styles. Lumpy stucco-and-sand plastered walls, half-timbering, thatched roofs, and sway-backed slate-roofed Norman "farms" and "stockbroker Tudor" appeared in the newly created garden suburbs. Authentic interiors and rustic antique furnishings completed the effect.

Figure 8-4 Log chalet, Camp Sagamore, Adirondacks, New York. At the turn of the century the wealthy sought out the "rustic life" in elaborate wilderness camps containing luxuriously appointed log structures.

Wood

Wood as an interior wall finish dates from the earliest primitive colonial shelters. As the pioneers moved westward, log structures were a natural result of the clearing of forested areas for farming. In western Pennsylvania and West Virginia many of these log structures continue to be inhabited, sometimes concealed on the outside by a later addition of clapboard. On the interior, the logs have a lime whitewash covering and mud-plaster chinking between. In rural areas during the nineteenth century, the exterior sheathing was exposed on the interiors of farm houses, barns, and outbuildings.

At the turn of the century wealthy families constructed elaborate rustic log "camps" in the Adirondack Mountains of New York State (Figure 8-4). These self-sufficient wilderness complexes contained many structures—boathouses, farm structures, stables, greenhouses, ice houses, cookhouses, laundries, recreation buildings, bowling alleys, hunting lodges, chalets, and guest cabins—linked by covered log walk-ways and porches. Huge rocks and boulders formed foundations, floors, and massive chimneys. The exposed log interiors were furnished with rustic twig furniture, and were decorated with animal trophies, fur throws, Indian blankets, and pottery. Also at the turn of the century, in Pasadena, California, the Green brothers developed their special, sophisticated interpretation of traditional Japanese wood construction combined with the decorative spirit of the Arts and Crafts Movement. These winter retreats for wealthy Midwesterners were even more luxurious than the great camps of the Adirondacks.

Following President Teddy Roosevelt's founding of the National Park Service in the early 1900s, rustic log park shelters and hotels were constructed in the West. The network of log extravaganzas continued to be expanded by the Works Progress Administration and Civilian Conservation Corps projects during the Depression years. Never before or since have any government recreational building programs been so sensitive to the ecology of wilderness areas. Many of these log structures are now more than 50 years old and require restoration and refurbishing.

Rustic wood structures have never lost their appeal. The Woodstock generation's "return to nature" brought a new demand for log buildings. Prefabricated log cabins are now available in kit form in New England and western states. Former President Jimmy Carter has retired to a hand-hewn log cabin in Georgia with rustic furniture he designed and built. So it is evident that skilled craftsmen can still be found in many parts of the country to restore the old log structures.

Many practical problems exist with both old and new log construction. The most serious one is fire because of the isolated rural location of most of these properties. Also, because of high fuel costs and safety considerations, these structures are not heated consistently, which leads to deterioration of the log structure, interior furnishings, and contents. Some experimental techniques are available for wood consolidation with chemical and plastic infusions.

Prior to the twentieth-century development of plywood, it was difficult to cover large interior surfaces with wood. The restricted width and linear character of sawn planks were constraints. The traditional skill of joinery was necessary to artfully fit the pieces together as paneling. Because wood is subject to dimensional variation related to humidity, paneling frequently shifts and may occasionally split or pull apart at the joints. This movement is particularly evident when paneled old doors or walls are painted. Cracking around the edges, exposing raw wood or previous color schemes, is a seasonal nuisance. New flexible acrylic caulking compounds seem to minimize some of these difficulties.

In the past, the wood chosen for paneling depended on what was plentiful and locally available. From the earliest colonial period, inexpensive paneling was usually made of pine and then painted. Solid oak was harder to work but was highly prized. Before the advent of plywood, fine furniture, paneling, and doors were made by gluing solid strips of wood together and then applying a veneered surface of an exotic or highly figured wood such as native tiger-eye maple, imported satinwood, or mahogany. During the nineteenth century, American interiors were resplendent with native walnut, black walnut, cherrywood, elm, and ash. During the late nineteenth century rich woods like rosewood and ebony were inlaid and combined with other exotic woods in marquetry work (Figure 8-5).

Up until the 1929 crash, whole European suites of old paneling were imported and incorporated in town houses and mansions. Fine English pine fireplaces, doorways, and staircases were skillfully dismantled, crated, and shipped along with antique furnishings, paintings, and decorative objects. If a building contains reassembled antique paneling, the same refitting techniques used to install the paneling can be employed to incorporate such modern conveniences as air conditioning, concealed television sets, and security equipment (Figure 8-6).

Tongue-and-groove boarding was created in the nineteenth century to solve the problem of warping and cupping. Some boarding was grooved or scored to give the impression of narrow strips. The "novelty" boarding was used as wainscot in utilitarian areas such as attics, kitchens, bathrooms, and railroad station waiting rooms. Originally it was darkly stained and then varnished; sometimes it was painted. This material can be easily duplicated by modern custom milling methods, and even mimicked in plywood.

Interior wood paneling is combustible and is not acceptable in most modern fire codes. The hazard may be somewhat mitigated by filling

Figure 8-5 Paneling, Villard Houses, New York, New York. In a gilded age, when barely an interior surface was left unadorned, fine wood paneling was embellished with marquetry borders and garlands in contrasting exotic woods, burls, or mother-of-pearl inlays.

Figure 8-6 Reproduction of wood carving. Although some types of traditional wood paneling and moldings may be reproduced by more economical modern techniques, most curved and sculpted decorative elements in high relief can only be duplicated by hand carving.

hollow spaces and voids behind the paneling with noncombustible foam or vermiculite. Applying fireproof coatings or pressure treatment may cause warpage and delamination of veneers, and may ruin the old patina and finish. In public spaces, such as lobbies, stairwells, and exit passages, it may be necessary to install automatic sprinkler systems.

Glass

Like masonry and wood, glass can be an architectural interior finish. It is also a purely decorative interior element. Originally an imported luxury, its use became more widespread as technology enabled cheaper production costs. (Glass tiles and structural glass facing are discussed in Chapter 7.)

Mosaics. Along with the nineteenth-century taste for historic revival styles there was an interest in adapting ancient inlaid mosaic techniques to contemporary needs. The traditional materials for mosaics are small chips of clay or stone occasionally accented with "gold" Venetian glass inserts. Venetian mosaic, composed entirely of glass tiles, is a technique notable for its luminosity and depth of color. Not only were Romanesque Revival churches, banks, and post offices embellished with mosaic floors, walls, and vaulted ceilings, but corner drugstores, saloons, and ice cream parlors also used a field of geometric mosaic tile. (For information on how to clean, repair, or replace mosaic see under "Ceramic and Mosaic Tile," earlier in this chapter.)

Transoms. Before electricity, every effort was made to direct daylight and air into corridors and interior spaces. Transoms over doors were not only glazed, but were frequently constructed to permit ventilation. Elaborately leaded, clear glass rectangles or semicircles formed transoms as well as sidelights in the typical Federal style entry halls. These decorative transoms are still intact but frequently painted over. Great caution must be used to avoid damaging the brittle leading when removing the paint. As in restoring stained glass, it may be necessary to replace leading that is bulging or falling apart.

Mirrors. Until modern times, fine-quality mirrors were silvered with mercury, which combined with the irregularities of old glass, to produce a special quality that is difficult to duplicate. Today the recognized health hazards of exposure to mercury have restricted its usage. Historically, mirrors were used not only to enhance the limited illumination of candlelit interiors but also to give the illusion of space or to extend by repetition a series of symmetrically placed windows on interior walls.

During the Art Deco period from 1920 to 1940, much use was made of tinted mirrors of gold, pink, and blue glass, including in many interior installations of fireplace surrounds and custom furniture. During the 1960s solar bronze and solar gray glass were added to the palette of decorative mirrors available to designers.

Decorative Glass. Decorative effects can be accomplished by controlling color, thickness, surface texture, and finish. Glass can be etched, sandblasted, or carved to create surface designs or patterns. If made opaque, backpainted or mirrored glass can be used as an applied decorative finish. Molded can be used as solid glass bricks or hollow blocks.

Specialized Painting and Decorating

Throughout history there has been an urge to decorate plain surfaces. Painted patterns often employ symbolic representations of natural forms, such as vines and flowers, as decorative flourishes. Sometimes paint is used to simulate unattainable rare or exotic materials such as wood burl or marble. These are called *faux* finishes.

Graining, Marbling, and Trompe l'Oeil. From the earliest colonial times humble American pine furniture and paneling were "grained" to look like mahogany or bird's-eye maple; wood columns and mantelpieces were "marbleized" to make them more elegant. Today we accomplish similar feats with plastic laminates such as Formica. Tabletops in humble diners and bathroom vanities simulate onyx, maple butcher block, and elm burl.

If ordinary folk aspired to grained and marbleized elegance, the well-to-do indulged themselves in even more grandiose fantasies by commissioning totally painted trompe l'oeil environments. The resulting murals depicted elaborate architectural spaces, carved details, statuary, imaginary vistas of elegant gardens, or idyllic classical landscapes with ruined temples and shepherds. By the nineteenth century hand-painted and wood block-printed scenic wallpapers were imported from Europe and trompe l'oeil became more common. Whereas most graining and marbling was executed directly on bare plaster, wood, or stone, "scenic" paintings were usually done on canvas and then applied to the walls, like wallpaper, with wheatpaste.

Restoration efforts are complicated when the effects of dampness have led to deterioration of the walls or surfaces on which the graining or marbleizing was applied. Museum-quality art conservation techniques are painstaking, slow, and impractical when they must be attempted far away from the well-equipped laboratory. Depending on the extent of the damage and the constraints of budget, it may not be feasible to restore original finishes. In many cases it is easier and cheaper to repaint the graining and marbling, by using the original techniques after the deteriorated surfaces have been repaired or replaced.

Today the increased historicism of contemporary architects and the growth of the postmodern aesthetic, which rejects the unadorned severity of the International style, has brought a revival of interest in architectural trompe l'oeil. Celebrated artists like Richard Haas have decorated New York City spaces as small as a SoHo loft in the manner of a Pompeiian villa and transformed the plain multistory former light court of the Alywnn Court Apartments into a "French Renaissance Revival" atrium that mimicks its own elaborate exterior carvings.

Stenciling. Stenciling as a technique for simple repeat decoration is an ancient one. In rural areas, in the colonial period, the use of stencils was less expensive than wallpaper. Much of this stenciling is prized today as folk art by collectors and museums of Pennsylvania Dutch, Shaker, and Scandinavian artifacts.

In the late nineteenth century, courthouse, theater, and office building interiors were decorated with painted stencil designs. Unlike the humble rural structures of earlier periods, these interiors contained complexly shaped, usually plaster, surfaces such as difficult-to-decorate vaults and domes (Figure 8-7). Some of these elaborate wall coverings were executed on canvas and, as in the case of Louis Sullivan's designs for the Trading Room of the Chicago Stock Exchange, required up to 75 different colors, including gold leaf.

The problems of restoring stenciling are similar to those of restoring graining and marbling and depend to a great extent on whether they were applied directly to bare surfaces or to removable canvas. If fragments of the original patterns have survived, new stencils may be cut and the colors matched to produce an accurate reproduction.

APPLIED WALL FINISHES

Finishes to walls that are removable are distinguished from the wall treatments just discussed, which are called architectural wall finishes.

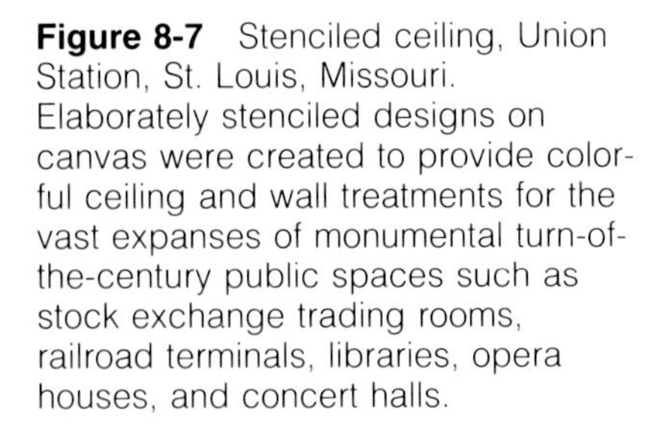

Figure 8-7 Stenciled ceiling, Union Station, St. Louis, Missouri. Elaborately stenciled designs on canvas were created to provide colorful ceiling and wall treatments for the vast expanses of monumental turn-of-the-century public spaces such as stock exchange trading rooms, railroad terminals, libraries, opera houses, and concert halls.

Wallpaper

Before the American Revolution and the establishment of a domestic industry, a large variety of imported wallpapers was available. During the eighteenth century, lead-based paints were expensive enough to make printed wallpapers competitive. Imported hand-painted wallpapers and scenic panels were also highly prized but much more expensive. Frequently during the course of renovating, which involves the removal of cabinets or wood trim, a hidden fragment of old wallpaper is revealed.

Finding exact wallpaper duplicates is not very likely and the cost of commissioning a reproduction in limited quantities is prohibitive. However, many period reproduction wallpapers are available today, so that unless the interior to be restored is of museum quality, an appropriate design of the same period is a good choice. Many museums and historic societies have established licensing arrangements with manufacturers for the reproduction of wallcoverings and textiles in their collections or their furnished period rooms. The Williamsburg, Winterthur, and Savannah collections, for example, provide special marketing caches for their products and royalties for the institutions. Sometimes these reproductions employ the original techniques of wood block and silk screen, but in most cases they are printed by automated modern methods.

To describe all the varieties of special wall coverings and their historic development is beyond the scope of this book. From the late nineteenth century to the present, wallcoverings of grass cloth, cork, laminated fabric, metal-like material, and vinyl are but a sampling. Today, many wall coverings are washable.

Fabric

Cloth wall coverings have always had a connotation of richness and comfort. During the medieval period, tapestry wall hangings were rolled up as part of the movable furnishings that were transported from castle to castle as the nobility traveled around their domains. Not only were tapestries works of art with complicated allegories, but they sealed against drafts.

Later, as royalty settled comfortably into their châteaus and palaces, many of these tapestries were permanently stretched on frames that were mounted to the walls and surrounded with carved wood paneling. Eventually, panels of richly patterned hand-cut velvets, silk brocades, and damasks imported from the Far East were sewn together and stretched onto the walls.

With the rise of prosperous merchants and landed gentry in the eighteenth century, the taste for cloth wall hangings was satisfied by the new process of copperplate engravings printed on cotton. Although typically decorated with vignettes of pastoral landscapes and romantic ruins, some of these monochromatic textiles (red, green, blue, brown, or black ink on white background) commemorated important events like hot-air ballooning ascents and the American Revolution.

During the nineteenth century, the newly rich merchants and industrialists adorned their town houses and country estates with textile wall coverings in sumptuous imported satins, silks, and jaquard-woven brocades. Heavy velvet hangings laden with clusters of silk fringe framed windows and formed portières at the doorways. The interiors of art galleries, museums, theaters, opera houses, and even private railway coaches were hung with heavy velours and furnished with plush mohair upholstery.

The traditional method of stretching fabric on nailer strips made of wood is vulnerable to extremes of humidity that cause sagging or tautness. Padding, which is often added to improve acoustical quality in auditoriums, compounds problems of fireproofing, cleaning, and maintenance. the use of Velcro simplifies stretching and removing fabric wall coverings while faithfully simulating the appearance of traditional installations.

Leather and Suede

In horse-and-buggy times, leather was much admired but not much used in interiors because of its cost and fragility. It was set in panels bordered with large brass nailheads and padded like a quilt as a covering material for doors. The leather was sometimes embossed with patterns or grains and stamped with gold leaf like fine bookbindings.

Richly tooled cordovan leather was favored by turn-of-the-century architects like Stanford White for such sedate preserves as private libraries, smoking rooms, and the senate chamber in Albany, New York, where it was a perfect accompaniment to overstuffed, leather-upholstered Chesterfield couches, lounge chairs, and banquettes of the period.

Suede tended to be used as a lining material for fine cabinetry and jewelry boxes, and also as an upholstery material despite its fragility and tendency to mark or stain easily. Both embossed leather and suede wall coverings have been superbly reproduced in vinyl, which is far more serviceable.

Lincrustia Walton

A popular turn-of-the-century wall covering material was Lincrustia Walton, a pressed linseed material that was available in sheet or roll form and could be surface-embossed to simulate patterns and textures resembling tooled cordovan leather (Figure 8-8). Lincrustia was capable of receiving various paint and stain finishes. Highly durable and serviceable, it was often used as a wall covering above the wood wainscot of stair halls and as a frieze in high-ceilinged late nineteenth-century parlors between the wood picture molding and the plaster ceiling cove. Because installations of Lincrustia in low-relief patterns were often combined with molded plaster detailing and then painted the same color, it is sometimes difficult to identify. Some original patterns are now in production again and are available from renovators' supply houses. Custom-embossed vinyl can now be made to reproduce original patterns, but it is costly.

Figure 8-8 Lincrustia Walton above a paneled wood dado, Ballantine House, Newark Museum, New Jersey. No self-respecting nouveau riche millionaire would live in a house with bare walls. Lavish wallcoverings, highly polished woodwork, gleaming brass hardware, decorative plaster ceilings, and stained glass were de rigueur. No surface was left unadorned.

Pressed Tin

Pressed tin was used originally as a fireproofing material on the ceilings of turn-of-the-century commercial spaces and meeting rooms. Embossed in a wide variety of patterns, it was relatively inexpensive and when painted looked like decorative plaster. Wall treatments also were executed in pressed tin. Old tin wall coverings have suffered damage by impact and have been cut to accommodate wiring, pipes, and ducts. Small missing sections can be reproduced from injection-molded plastic by making a cast. Recently, pressed tinplated ceilings have made a comeback, so many traditional patterns are now available.

DECORATIVE METALWORK AND HARDWARE

During the colonial period, the primary materials for decorative hardware were wrought-iron and brass. Most hardware was functional and striking in its simplicity. All elements were hand-wrought and finished by local blacksmiths. Door and shutter hardware, fireplace fittings, hooks, lamp brackets, and an occasional railing were about the extent of available wrought-iron items. Chased brass doorknobs, box locks, and door knockers were seen with name-engraved plaques.

The romantic revival styles demanded more elaborate metal hardware and fittings. The Gothic Revival aesthetic was a distinct departure from the elegant simplicity of Georgian, Federal, and Classic Revival style hardware. As mid-nineteenth-century architectural styles and interior decor became more ornate, so did the hardware, which favored sculpted, swirling motifs of garlands, flowers, animals, and grotesque masks.

Figure 8-9 Plaster ceiling, Empire Theater, New York. Cast plaster decorative elements, moldings, medallions and cartouches on a domed theater ceiling. Missing or damaged elements can be easily duplicated by casting from the remaining original portions. New air conditioning grilles have been placed in a relatively unobtrusive manner.

The Art Deco style relied heavily on decorative metal elements—wrought-iron grillwork, screens, railings, balconies, furniture elements, and fireplace screens and equipment. Also popular were multicolored metallic assemblages of bronze, copper, and brass used on elevator doors and clocks, numerals, signs, architectural letterings, radiator grilles, and as lobby ornaments in the new skyscrapers.

Much of the fine decorative metalwork of all these periods and styles has outlived the structures and interiors for which it was originally created. Flea markets, parts warehouses, and antique dealers are a good source for locating restoration items.

CEILINGS

In the earliest wood-framed American colonial structures, a ceiling was the exposed undersurface of the wood floor or roof above. With the passage of time and the greater availability of milled timber, much of the exposed rough framing was concealed by encasing the heavy timber posts and beams. Later, ceilings were plastered so that the floorboards were no longer visible. These early plaster ceilings were completely plain and whitewashed.

During the late seventeenth and early eighteenth centuries some well-to-do Dutch patroons, Southern plantation owners, and urban merchants commissioned ornamental plaster ceilings carved in the European fashion with robust garlands of flowers and fruit. These elaborate installations were usually limited to the principal entertaining rooms and public spaces. Of course, most interiors of this early period have plain ceilings that are easily restored or reproduced.

Ceilings with exposed beams or framing present greater difficulties for concealing modern electrical, communications, and security-alarm wiring; plumbing; sprinklers; and ducts. Alternative locations may often require complex installations in cellars, attics, and unutilized fireplaces and chimneys.

Old plaster ceilings must be replaced when cracked or loose portions separate from the wood lath beneath. (Patching plaster ceilings is discussed under "Plaster," in Chapter 7.)

Ornamental Plaster Ceilings

The eclectic parade of historic revival styles brought replicas of ancient and exotic architecture to the American landscape. Certain adaptations were inevitable, particularly in the choice of building methods. The classic detailing of the Greek Revival and the splendors of Egyptian Revival were executed in wood and plaster rather than the marble and stone of antiquity.

Even though the outside of most Gothic Revival churches and country estates was faced with stone, the vaulted interior ceilings and carved decorations were made of wood and plaster artfully scored and painted to simulate the masonry joints of the originals.

The mid-nineteenth century was characterized by the lavishness of the fashionable Anglo-Italianate and French Renaissance Revival styles. Some of this heavily sculpted baroque plaster ceiling ornament was so three-dimensional that freestanding elements such as leaves, pierced fretworks, finials, and drops required molds as complex as those used to cast heroic sculpture.

These overhead extravaganzas—ceiling paintings on canvas set in gold-leafed, carved decorative plaster panels—reached their zenith in the grandiose eclectic interiors of the Beaux-Arts style at the turn of the century. Many of our distinguished monumental landmarks and public buildings are of this era. Today ceilings of customs houses, city halls, museums, libraries, theaters, and concert halls are being restored to their former grandeur (Figure 8-9).

The complex design of monumental Beaux-Arts style structures was such that the outside structural envelope, with its domes and pedi-

ments, did not directly reflect the forms or plan of the interior spaces. A completely independent structure formed the intricately shaped interior spaces. Following baroque precedents, it was common to have double domes (Figure 8-10). The external copper or lead-sheathed dome covered the smaller, more intimately scaled interior plaster dome, which was formed over a vault of hollow terra cotta blocks or an iron armature.

At the turn of the century, because dependable artificial lighting had not yet replaced daylight, public spaces and even residential interiors contained large decorative skylight panels of tinted or stained glass. The ornamental glass panels were shielded above from the weather by clear glass lanterns resembling iron-framed miniature greenhouses. Frequently these exterior lanterns were the source of water penetration and leaks that caused major damage to the plaster ceilings in which the decorative glass skylight panels were mounted.

Restoring Ornamental Ceilings. Restoring plaster ceilings with elaborate ornamentation is much more complicated than restoring undecorated ones (Figure 8-11). Often the damage occurs to the plain field of the ceiling rather than to the decorative borders, cornices, or central medallion. If this is the case, it may be possible to leave the decorative elements intact and repair or replace only the damaged field portions. If it appears that the decorative elements may be damaged in the process of repairing the ceiling, or that major portions are already missing, it might be better to make molds of typical sections prior to any construction. (It is much easier and cheaper to make new casts of the original elements than it is to carve new models if they are lost.) Unlike the reproduction of terra cotta, where clay shrinkage must be reckoned in the casting process, direct molds can be made for plaster replacement elements.

Figure 8-10 Old stone bank, Providence, Rhode Island. Cast-iron framed domes and cupolas formed the exterior silhouettes of major nineteenth-century buildings and also provided a method of skylighting dark interior spaces in the pre-electric-illumination era.

Figure 8-11 Ceiling restoration, Drayton Hall, South Carolina. Wealthy Southern planters attempted to reproduce fashionable English Georgian eighteenth-century decorative style in their lavish manor houses. Two centuries later, these heavily carved plaster ceilings frequently require painstaking restoration. The floor boards above the ceiling have been lifted so that the plaster ceiling can be resecured to the joists.

Figure 8-12 Pressed tin ceiling, Chenango County Courthouse, Norwich, New York. In the late nineteenth century, pressed tin was more than just a decorative fashion, it provided both increased fire resistance and a cheap form of decoration.

Figure 8-13 False ceiling, Garvan/ Brady House, Albany, New York. A false ceiling was discovered when restoration work was begun to adapt an elegant town house to a government studies center.

Figure 8-14 Restored dome, Garvan/ Brady House, Albany, New York. Removal of a false plaster ceiling revealed a long-forgotten fresco painted on a domed upper story.

WOOD

Many late nineteenth-century public buildings with large assembly spaces, such as grange and music halls, churches, schools, hotel dining rooms, ballrooms, casinos, or railroad stations, were spanned by wood trusses and wood-framed roofs and ceilings. The ceiling surfaces were commonly covered with tongue-and-groove flush boarding or novelty boarding (decorative scoring or beading giving the impression of narrow strips), which was varnished or painted and stenciled.

These ceilings bulged or sagged as a result of leaking roofs. Restoration of these structures is restricted by modern building code compliance, which requires stringent fire protection in areas of public assembly. In order to retain the original character of the structure, and restore the damaged wood ceilings, installing costly concealed fire-detection or sprinkler systems may be an unavoidable necessity.

PRESSED TIN

Disastrous fires in the nineteenth century paved the way for modern fire and building codes. An early recognition of the need for fire-resistant construction was the use of pressed tin as a ceiling treatment. While originally intended as a fireproofing measure, its great popularity rested on its low cost and its convincing resemblance to decorative plaster (Figure 8-12). The main public hearing room in the old city hall in Corning, New York, is entirely covered—walls and ceilings—with pressed tin ornamentation.

The recent wave of nostalgia for the Gay Nineties and ceiling fans seems to have prodded the moribund pressed tin industry into new vitality. Many of the traditional patterns are now available and usually a reasonable facsimile can be found. If an exact replica is necessary to replace missing elements it is possible to make an inexpensive thin plastic reproduction by the injection-molding process. The only drawback is that the decorative reproduction may not be acceptable because it is not fireproof.

SUSPENDED CEILINGS AND ACOUSTICAL TILES

During the immediate post-World War II period, which was strong on urban renewal, much of our architectural heritage was razed. Too much of what escaped the bulldozer was modernized beyond recognition. Today, as we seek to revitalize our downtowns, we are taking a new look at run-down old movie palaces, office structures, and public buildings. On closer examination, many of these faded modernizations are more superficial than one might expect. By carefully removing suspended false ceilings and pasted-on acoustical tile, we are often rewarded with the discovery of the virtually intact original ceiling. By patiently replacing missing elements and patching damaged portions, these once-grand spaces are restored to their original luster and attractiveness (see Figures 8-13 and 8-14).

VIEWPOINTS ON FURNISHING THE HISTORIC INTERIOR

There is a general consensus on the appropriate restoration of the exterior of an old building; designing the interior is more problematic. In the most significant buildings, in all styles and periods, the architect has played a key role in determining the character and detailing of the principal interior spaces, often designing or selecting the furnishings as part of the ensemble.

The role of the designer or decorator as independent professional in the design process evolved at the turn of the century. In the tradition of the Edwardian tastemakers, Edith Wharton, together with architect Ogden Codman, wrote a book on the decoration of houses; Elsie de Wolfe followed. De Wolfe established herself as a professional decorator. Gradually a schism developed between those preferring traditional design and those championing the modernist approach. Today interior designers, just as architects, tend to specialize in either traditional or modern approaches in design. The growing interest in historic preservation, as well as the increased influence of historicism in architecture, is leading many architects and interior designers to develop a more eclectic approach.

It is becoming more common for designers to restore the architectural interior shell to its traditional appearance, regardless of whether it will be decorated with appropriate period antiques or serve as a contrasting background for modern furnishings. Now that pioneering twentieth-century furniture has become classic, prime examples are collected and displayed with the same pride as the more traditional antiques. In its American Wing, the Metropolitan Museum of Art in New York has a completely restored and furnished living room removed from an exceptional early twentieth-century Frank Lloyd Wright-designed house in Minnesota. More fortunately, Frank Lloyd Wright's renowned Fallingwater in Pennsylvania now functions as a house museum and its original surroundings have been transformed into a vast nature preserve operated by the Western Pennsylvania Conservancy.

The Green Brothers' Gamble House in Pasadena, California, and Walter Gropius's house in Lincoln, Massachusetts, are being preserved as house museums in their original settings; and some Richard Neutra and Irving Gill houses in southern California are also being restored. The curators of these modern house museums are as scrupulous about the authenticity of every rug, object, painting, and piece of furniture as they would be in restoring a seventeenth-century colonial saltbox.

Historic houses are still livable today and may incorporate all the improvements of modern technology. The same cannot be said of certain functionally obsolete nonresidential structures, such as steamship piers, railroad terminals, and movie palaces. Even those buildings whose original function is maintained, such as banks, post offices, and libraries, have been transformed by automation and the computer revolution. Adaptation to present-day requirements if often at odds with the traditional arrangement of the major public spaces these buildings contain. Beaux-Arts monumentality is not always compatible with automated tellers, postage machines, coin-operated copiers, and television security systems.

Extraordinary interiors, such as Louis Sullivan's Auditorium Building in Chicago, New York's Radio City Music Hall, and Frank Lloyd Wright's Johnson Wax Company headquarters in Racine, Wisconsin, have required tremendous preservation efforts to extend their functional use without compromising their original design.

The original designers of most monumental public buildings rarely, if ever, took into account the needs of the elderly or handicapped The introduction of ramps, special exiting accommodations, and other aids for the disabled is a formidable challenge to the preservationist (Figure 8-15).

Figure 8-15 Jefferson Market Library, New York, New York. The conversion of this 1880s High Victorian Gothic courthouse into a local library required several internal modifications to satisfy building code and safety requirements. The contemporary "bridge" across the main reading room, formerly the courtroom, functions as a fire passage linking a mezzanine to a secondary means of egress.

Restoration techniques for early modern interiors are essentially the same as for traditional structures, with the exception that some new and experimental materials were used in these buildings. Some of these once-new materials and construction techniques have not stood the test of time and have caused problems of their own. Frank Lloyd Wright's use of Pyrex glass tubing as clerestories in the Johnson Wax headquarters was a total failure. The subsequent development of plastic tubing has made it possible to achieve the architect's original intent and solve the nuisance of thermal leakage and water penetration, problems that were damaging the interior spaces and furnishings.

Similarly, experimental technology has resulted in the need to replace the deteriorated radiant-heat piping system beneath the waxed herringbone brick floor of Phililp Johnson's Glass House in New Canaan, Connecticut. The restoration problems of early modern buildings may be very different from those of traditional structures, but no less challenging to remedy.

CHAPTER NINE

INTEGRATING MODERN NECESSITIES

TRADE-OFFS, LIMITS, STRATEGIES

Preserving and maintaining old structures engenders complications unforeseen by the original builders. Restored landmarks are expected to provide not only authenticity but cost effectiveness, energy efficiency, and modern conveniences as well. The unobtrusive installation of modern plumbing, electrical systems, lighting, communications, climatization, and security systems poses a far greater challenge to the architect and preservationist than just a literal restoration or rehabilitation.

INCREASING COMFORT AND ENERGY EFFICIENCY

Much of the research on designing for maximum energy efficiency is based on new construction, where optimum conditions are controllable. Trying to accomplish the same goals with old buildings is not only impractical but also highly undesirable. In an old building, achieving maximum energy conservation, in the modern sense, can cause the permanent loss of character and historic integrity.

Some compromises, of course, are reasonable and necessary, particularly if a sensitive approach to historic preservation is to be extended beyond the special category of major monuments and museum restorations, into the mainstream recycling of older neighborhoods and suburban communities (Figure 9-1). In most historic district legislation, only the portions of a building visible from a public way are regulated. However, the individual owner has the right to do what he or she wishes with the remaining elevations and the interior. Obviously, it is easier and less costly to achieve greater energy savings when only portions of the exterior are restricted.

Figure 9-1 Air conditioning units, brownstone row house, Brooklyn, New York. Until technological advances provide a more satisfactory solution, the most practical method of air conditioning a traditional row house is with individual window or through-the-wall units. Their location must be carefully chosen to minimize the disruption of decorative features and architectural detailing on the interior as well as the exterior.

Analyzing New Needs for Old Buildings

An energy audit or analysis should be conducted to determine what steps are needed to make an old building more energy efficient. Some local utility companies offer these services for residential customers, but the personnel who conduct these surveys are seldom experienced in dealing with the special problems of historic structures. Some of their conventional advice can, in fact, be quite threatening to the appearance of a significant architectural landmark.

A good energy conservation analysis must clearly distinguish between the implications of the construction of the building's exterior envelope and those demands imposed by the spaces and programmatic functions it contains. Climate, exposure, and siting are all significant determinants of energy requirements regardless of the type of construction or the interior functional requirements.

To provide the most thorough analysis, a study team including the architect, a mechanical engineer, a representative of the owner, the manager of the facility, and operating or maintenance personnel should be assembled. This is the optimum arrangement and is usually possible when an existing setting is being studied for extended use. However, in adaptive-use projects the managing and operating staff has seldom been hired at the critical early planning stage. Under these circumstances, the professional consultants must assume the responsibility for planning many of the operational aspects. This does not always result in maximum efficiency and may require modifications after the facility is in actual operation.

The first task of the study team is to survey all existing facilities and document mechanical systems and patterns of usage. Special requirements and items of unsatisfactory performance should be noted. Observations of existing energy consumption should be made, with projected future demands part of the study. When greater operating efficiency is an important objective, these visits and observations must be patiently carried out at different times of the day and different days of the week to fully note patterns of use and varying energy demands.

Once this survey is completed, it may be possible for the planning team to recommend primary operational changes that will require minimal physical intrusion. Some economies can be made, for example, merely by reducing wattage of lighting, putting timers on mechanical systems, or installing a new thermostat capable of day and night settings. These easy steps may yield modest savings, but more significant economies may require substantial capital expenditures for physical improvements and replacement of outdated and inefficient mechanical systems. Although the long-term benefits may justify the investment, it may be difficult to finance them all at once. The planning team must establish realistic goals and priorities so that a phased program of improvements can be implemented. A cost-benefit analysis may demonstrate, for example, that the fuel costs saved by replacing an inefficient furnace may be a more prudent first step than increasing the insulation value of walls and ceilings.

A truly comprehensive energy-saving strategy for an old building takes into account the mechanical and energy-consuming equipment; the people who own, operate, and use the building; and the special requirements of the functions that the building has.

INTRODUCING MODERN CLIMATIZATION SYSTEMS AND MEASURES

Introducing modern climatization systems must be carefully considered. Many old buildings had climatization systems that were relatively primitive compared to our modern concept of comfort. Most of these systems were incapable of producing consistent, controllable heating or cooling levels. Many early heating systems were extremely unpredictable and hazardous. Fireplaces and stoves cannot always be successfully combined with central air heating and cooling systems. Particular caution must be exercised when installing new energy-saving insulation and double glazing because both radically diminish the amount of fresh air that flows into a building (see Figure 9-2).

Modern environmental systems that regulate temperature and humidity can seriously upset the balance of a structure accustomed to more gradual transitions. The installation of energy-saving heavy insulation and storm windows can trap condensation in wood-framed exterior walls. (See also the sections on "Wood" and "Mechanical Systems" in Chapter 7.)

Increasingly, private and institutional owners are concerned not only with the preservation of the exterior but also with distinct interior spaces. Frequently, when old buildings are being restored for extended or similar uses, it is possible to limit the initial investment by upgrading rather than totally replacing the existing mechanical systems. This approach is much less disruptive to the traditional interior because no new mechanical appurtenances such as ceiling diffuser grilles or convector units are needed. Many efficiencies can be achieved unobtru-

Figure 9-2 Traditional bank vestibule, Canandaigua, New York. In order to increase the energy efficiency of a restored bank on a typical small town commercial block, a new interior vestibule was designed of marble and traditional architectural bronze components.

sively in the boiler room by reducing the fuel consumption of the furnace and providing a more sophisticated thermostatic control system.

However, the upgrade approach is not always feasible. Many adaptive-use solutions require such dramatic rearrangements of the interior spaces that the upgrading of the old mechanical systems requires too many extensive changes to be economical. When enormous single spaces, such as churches and theaters, are portioned into smaller units, the existing heating, ventilating, and cooling systems are not salvageable.

Insulation

Ever since the energy crisis of the 1970s, there has been greater emphasis on installing insulation to provide fuel economies. Some older buildings already have some form of insulation and this should be taken into account before proposing a complete new installation. Early builders packed hollow wall cavities with organic materials, such as seaweed, animal hair, and clay, depending on the region. By the beginning of the twentieth century, mica chips, cellulose fibers, mineral wool-batts, and asbestos were commonly used as insulating materials. Many of these materials have proven to be combustible and also to produce toxic fumes. Under damp conditions, many of the organic materials deteriorate and attract pest infestation. Generally, it is better to replace organic materials with more inert, modern fireproof insulation.

Sealing a building for energy efficiency often has harmful side effects. A tightly sealed building traps condensation in the exterior wall construction and risks serious structural damage from dry rot. Scientists have discovered that snugly sealed buildings are not healthy because many contaminants that would normally be dispersed are trapped. Serious health problems are linked to breathing fine particles of asbestos and to toxic fumes from urethane foam insulation. In many circumstances, it may be necessary to remove and replace these installations.

Exterior Siding

Owners of old buildings are often tempted to find ways to minimize the high cost of maintenance and upkeep. Prompted by the constant nuisance of painting and repairing wood structures, many owners have installed exterior siding. Installing effective modern insulation requires a vapor barrier on the warm side and ventilation on the cool side to allow condensation to evaporate. Most old buildings have no vapor barriers, and this condition is greatly exacerbated when metal or vinyl siding is applied to the outside. As with new insulation, condensation can-

Figure 9-3 Old courtroom, City Hall, Oswego, New York. Prior to restoration a dropped modern hung ceiling rested on top of the decorative window heads. The room was illuminated by unattractive stem-mounted fluorescent light fixtures.

not escape and accumulates within the exterior walls, causing serious structural damage from dry rot.

Applying an envelope of siding can also compound the difficulties of detecting and controlling the spread of fire. Many vinyl siding products, although inflammable, can contribute to loss of life by producing choking toxic fumes.

The removal of exterior siding during restoration may lead to the discovery of hidden architectural detailing, but more likely it will reveal deteriorated wood siding that will need to be replaced.

Climate Controls

Most thermostatically controlled interior climate systems are activated by a response to exterior temperature changes. In public and institutional buildings, levels of usage and occupancy may vary considerably. Providing comfortable conditions for the occupants may prove less than ideal for the preservation of a recycled old structure and its furnishings. Sophisticated climate control systems, which can maintain a fixed balance of temperature and humidity, are expensive to install and maintain. The most common heating methods used at the turn of the century were gravity heat (ducted warm air) and steam heat, neither of which contained any provision for replacing the moisture lost from the steady heating of interior spaces. It is not always physically or economically possible to upgrade old heating systems to modern comfort standards. It is possible to regulate air temperature in a gravity system. However, a steam system must be maintained at a constant temperature and is very inflexible and prone to overheating. (The section on "Mechanical Systems" in Chapter 7 discusses some practical approaches to climate control systems.)

LIGHTING

Most of us have been so accustomed to modern lighting that it is difficult to conceive of the dimness of traditional interiors. The traditional lighting systems of most old buildings are totally inadequate by modern standards and expectations. The successful resolution of these shortcomings is a matter of finding a balance between the current programmatic requirements of an old building and the original architectural character of its interior spaces. Even in museum restorations, where functional requirements are not as rigorous, it is difficult to adequately see the furnishings and decorative objects without some accent lighting to supplement the dim output of the original fixtures.

The most attractive method is to relamp the existing fixtures to pro-

Figure 9-4 Restored courtroom, City Hall, Oswego, New York. The hung ceiling has been removed, revealing the original decorative plaster ceiling. Electrified pendant, reproduction gasoliers provide a more authentic and attractive means of illumination.

vide more light, but this is usually insufficient. If similar old fixtures or reproductions are available, it may be possible to increase the number of ceiling fixtures or wall sconces consistent with the period character of the room (see Figure 9-3 and 9-4). If additional lighting is still needed, it is sometimes possible to conceal the new lighting within architectural and decorative elements such as ceiling moldings, valance boards, and niches. To achieve the desired lighting, it may be necessary to combine the relamped existing fixtures with supplemental concealed fixtures.

Large institutional and commercial installations require considerably more ingenious solutions to upgrading lighting systems. Providing the high lighting levels required for live television broadcasting has necessitated the installation of separate special lighting systems in old government buildings, courtrooms, and hotels. The cavernous spaces of monumental Beaux-Arts interiors often lend themselves to simpler lighting solutions than do smaller more austere spaces. The ornate ceilings and interior decor make it easier to conceal supplemental lighting.

Until the twentieth century, night illumination of major buildings was virtually impossible. Before electric lighting, clusters of lanterns illuminated by burning oil, candles, or gas were all that was available. None of these earlier forms of illumination could be easily directed as beams or shafts of light. At the turn of the century, the first form of modern exterior night lighting was the use of rows of small bare bulbs to light up theater marquees. Now that landmark buildings of the early twentieth century are being restored for extended and adaptive use, it is important to become familiar with the evolution of contemporary lighting.

Art Deco style buildings in the 1920s were the first buildings to incorporate modern lighting into the architectural design of their facades (Figure 9-5). Improved technology made the use of large areas of glass, mirrors, and glass block more feasible, increasing the translucency of illumination. The development of fluorescent lighting permitted a continuous linear source of light for the first time. After World War II the development of boxlike, sleek glass curtain walls eliminated conventional exterior architectural night lighting by substituting illumination from within.

Figure 9-5 Art Deco entrance, Daily News Building, New York, New York. In the 1920s electric illumination became an element of architectural design on the exterior as well as the interior. These early experiments included recessed, indirect, and cove lighting. Restoration of lighting fixtures of this period may require some updating or replacement with more modern wiring and lamps.

SECURITY SYSTEMS

With the incidence of vandalism and theft on the rise, property owners are concerned with upgrading their security. The requirements of historic structures demand special attention. Generally this implies the

Figure 9-6 Roll-up security gate, Greenwich Village, New York. Most urban merchants require greater nighttime security than old buildings originally provided. Combining the overhead protective gate housing with the store's signs minimizes the disfigurement these security installations have on old facades.

addition of protective metal grilles, gates, or rolling shutters at ground-floor windows and doorways, increased exterior lighting, and alarm systems (Figure 9-6). Unless the installations are carefully studied, applying conventional protective devices can mar the appearance of an architecturally significant landmark.

The problems of designing security systems for old buildings vary considerably, depending on the building's of construction, layout, and location. Generally, urban masonry structures, which abut each other or share common party walls, are easier to protect than isolated free-standing buildings where it is more difficult to restrict access.

Even though the risk of theft and vandalism may be greater in the city than the country, isolated rural buildings can be quite vulnerable. Some stylistically correct physical barriers, such as traditional wrought-iron gates and bars, may be unobtrusively fitted to an urban masonry structure but will look highly inappropriate on a country wood building. It may therefore be necessary to use a concealed electronic system to preserve the original exterior character.

One of the additional difficulties of providing a security system in an old building is the need to make it as unobtrusive as possible. Even if all the wiring is concealed, the sensing devices cannot be hidden. It may be necessary to provide more sophisticated security systems in order to provide adequate concealment of these devices. To plan the location of the sensing devices, it is essential to have some knowledge of the interior finishes and furnishings layout. This is particularly critical with electronic systems that depend on an unobstructed beam of light in order to function.

False alarms are a nuisance and a hazard. If a system is too elaborate, awkward, and inconvenient it will not be used effectively. Have the system zoned so that maintenance or supervisory personnel can have access to restricted portions of the structure to perform routine tasks without leaving the entire building vulnerable. No amount of electronic wizardry is a substitute for human observation and judgment in an emergency.

FIRE ALARM SYSTEMS

The first consideration of a fire alarm system is to provide prompt and accurate fire detection. Providing a safer, direct exit for people is the next priority. The primary focus of most building codes and fire rating systems is life-safety. With historic buildings, salvaging the irreplaceable landmark itself needs to be considered. This introduces separate concerns that are often at variance with the life-safety codes. Simply stated, optimum people protection may not be possible without compromising the historic and architectural character of the buildings. Compromises are not easily made by local fire and building inspectors whose discretionary powers are limited by statute. Ignoring modern safety provisions may also invalidate insurance coverage.

The risks are certainly high for a historic building constructed by traditional methods with materials that are not fireproof. Avoid adding any more combustible materials in the process of restoration and rehabilitation. Some substitutions are less noticeable than others. Patching with modern plaster over expanded metal mesh instead of traditional wood lath can duplicate the original texture of old plaster and still reduce the fire hazard. Many inexpensive and serviceable modern materials can simulate traditional finishes quite effectively. Extruded anodized aluminum window and storefront elements can reproduce bronze and other metals without the weight, necessary rustproofing, and periodic painting. Lexan plastic glazing is shatterproof and permits historic buildings to retain their original small-paned window character without sacrificing safety and security. Many modern paint sealers and coatings are technically superior to their traditional counterparts and give about the same appearance. Products like polyurethane floor sealers, also fireproof, have increased the wearability of old wood floors and reduced waxing and polishing. Most preservationists are willing to adopt some of the new and innovative materials and techniques.

Figure 9-7 New elevators, City Hall, Oswego, New York. In order to provide mandated access for the handicapped directly from the street level to all public areas of the building, a new elevator core of contemporary design, clad in matching stone, was constructed at the rear of the old city hall.

ACCESS FOR THE DISABLED

One special concern that was rarely addressed in old buildings was the special needs of the aged and disabled. Trying to accommodate these special requirements in historic structures is additionally difficult because of the desire to preserve the original architectural features. Ultimately some compromises must be made (Figures 9-7 and 9-8). Sometimes total accessibility means providing ramps or several elevators and is unachievable. Complex structure with many levels, towers, and nooks and crannies may not be suitable for public or institutional uses mandated to provide access for the handicapped. Many old mansions formerly used as schools, nursing homes, and conference centers are being abadoned by owners faced with the costs of complying with access requirements for the disabled.

Figure 9-8 Typical elevator lobby, City Hall, Oswego, New York. A glazed "notch" link encloses the new elevator lobbies and abuts the old mansard roof at the top floor level.

MAINTENANCE AND ROUTINE CARE

Restoration may make up for years of neglect, but the job is not complete unless a program for ongoing maintenance is worked out and followed. Establish a written schedule or manual of routine maintenance. This manual should outline the recommended procedures and the intervals at which each should be performed.

Storm damage and unexpected leaks cannot be anticipated, but many conditions can be. Owners should keep an up-to-date log of repairs and note all deteriorated conditions. Duplicate sets of drawings, specifications, and equipment manuals should be kept in a secure location for service and maintenance personnel and, of course, for future times. Preservation is not, after all, a one-time effort.

BIBLIOGRAPHY

BOOKS AND PAMPHLETS

ARCHITECTURAL HISTORY: COMPENDIUMS OF STYLES, BUILDING TYPES, AND CONSTRUCTION METHODS

Andrews, Wayne. *Architecture, Ambition, and Americans: A Social History of American Architecture.* Rev. ed. New York: Free Press, Macmillan, 1978. 334 pp., illus., biblio., index.
(a lively and informative account of the interaction that produced America's finest architecture)

Blumenson, John J.G. *Identifying American Architecture: A Pictorial Guide to Styles and Terms, 1600–1945.* Rev. ed. New York: W. W. Norton, 1981. 120 pp., illus., gloss., biblio.

Cole, Katherine H. and H. Ward Jandl. *Houses by Mail: A Field Guide to Mail-Order Houses from Sears, Roebuck and Company.* Washington, D.C.: Preservation Press, 1986. 400 pp., illus.
(a compendium of mass-produced houses available in the early twentieth century)

Condit, Carl. W. *American Building: Materials and Techniques from the Beginning of the Colonial Settlements to the Present.* 2nd ed. Chicago: University of Chicago Press, 1982. 342 pp., illus., biblio.

Foley, Mary Mix. *The American House.* New York: Harper & Row, 1981. 300 pp., illus., index.

Grow, Lawrence. *Classic Old House Plans: Three Centuries of American Domestic Architecture.* Pittstown, N.J.: Main Street Press, 1984. 128 pp., illus.

Jandl, H. Ward, ed. *The Technology of Historic American Buildings.* Washington, D.C.: Foundation for Preservation Technology, 1984. 225 pp., illus., index.
(subjects include Chicago balloon frame, I-beam, cast iron, terra cotta, iron builders' hardware, metal roofing, and nineteenth-century exterior painting)

Kaufmann, Edgar, Jr. *The Rise of an American Architecture.* New York: Praeger, 1970. 241 pp., illus., biblio.
(of particular interest is an article by Winston Wiseman, offering a new view of skyscraper history)

Maddex, Diane, ed. National Trust for Historic Preservation. *Built in the U.S.A.: American Buildings from Airports to Zoos.* Washington D.C.: Preservation Press, 1985. 192 pp., illus., biblio.

Pevsner, Nikolaus. *A History of Building Types.* Princeton:Princeton University Press, 1976. 352 pp., illus., biblio., index. (contains more English and European prototypes than American ones, but there is no domestic equivalent; an academic reference work)

Poppeliers, John, S. Allen Chambers, and Nancy B. Schwartz. *What Style Is It? A Guide to American Architecture.* Rev. ed. Washington, D.C.: Preservation Press, 1984. 112 pp., illus., gloss., biblio.
(a basic, concise guide)

Rifkind, Carole. *A Field Guide to American Architecture.* New York: New American Library, 1980. 336 pp., illus., biblio., index.
(profusely illustrated)

Roth, Leland M. *America Builds: Source Documents in American Architecture and Planning.* New York: Harper & Row, 1983. 666 pp., illus., index.

———. *A Concise History of American Architecture.* New York: Harper & Row, 1979. 400 pp., illus., gloss., biblio., index.
(a very good account)

Smith, G.E. Kidder. *Architecture in America: A Pictorial History.* New York: American Heritage, 1976. 2 vols., 832 pp., illus., index.

Walker, Lester. *American Shelter: An Illustrated Encyclopedia of the American Home.* New York: Overlook Press, Viking, 1981. 320 pp., illus., biblio., gloss., index.

Whiffen, Marcus. *American Architecture Since 1780: A Guide to Styles.* Cambridge: MIT Press, 1969. 328 pp., illus., gloss., index.
(an excellent well-illustrated, portable, and practical guide)

——— and Frederick Koeper. *American Architecture, 1607–1976.* Cambridge: MIT Press, 1981. 2 vols., 576 pp., illus., biblio., index.
(an excellent reference book)

ARCHITECTURAL HISTORY: INDIVIDUAL PERIODS, REGIONAL STYLES, AND ARCHITECTS

Baker, Paul R. *Richard Morris Hunt.* Cambridge: MIT Press, 1980. 588 pp., illus.
(one of the first American architects to be educated at the École des Beaux Arts, Hunt was extremely influential in the late nineteenth century)

Brocks, H. Allen. *The Prairie School: Frank Lloyd Wright and His Midwest Contemporaries.* Toronto: University of Toronto Press, 1972. 373 pp., illus., biblio., index. (Reprint ed. New York: W.W. Norton, 1976.)

Carrott, Richard G. *The Egyptian Revival: Its Sources, Monuments and Meaning, 1808–1858.* Berkeley: University of California Press, 1978. 221 pp., illus., biblio.
(documents an odd but unique style)

Cerwinske, Laura. *Tropical Deco: The Architecture and Design of Old Miami Beach.* New York: Rizzoli, 1981. 96 pp., illus.

Cummings, Abbott Lowell. *The Framed Houses of Massachusetts Bay, 1625–1725.* Cambridge: Belknap Press, Harvard University Press, 1979. 261 pp., illus., index.
(a basic reference pertaining to early buildings in New England)

Curl, Donald. *Mizner's Florida: American Resort Architecture.* Cambridge: MIT Press, 1984. 256 pp., illus.
(the master of the Mediterranean style, Mizner was the architect and developer of Palm Beach and Boca Raton)

Cuthbert, John, Barry Ward, and Maggie Keeler. *Vernacular Architecture in America: A Selective Bibliography.* Boston: G.K. Hall, 1985. 145 pp., illus., index.

Downing, Andrew Jackson. *Victorian Cottage Residences.* Reprint ed. New York: Dover, 1981. 352 pp., illus.

Gebhard, David and Harriette Von Breton. *L.A. in the Thirties.* Salt Lake City: Peregrine Smith, 1975. 165 pp., illus., biblio.

Girouard, Mark. *Sweetness and Light: The Queen Anne Movement, 1860–1900.* New Haven: Yale University Press, 1984. 250 pp., illus., index.
(an account of the popular English style that greatly influenced American architecture after the Centennial Exposition held in Philadelphia in 1876)

Greiff, Constance M. *John Notman, Architect, 1810–1865.* Philadelphia: The Athenaeum, 1979. 253 pp., illus., biblio., index.
(covers one of America's most innovative and influential nineteenth-century architects)

Grow, Lawrence and Dina Von Zweck. *American Victorian: A Style and Source Book.* New York: Harper & Row, 1984. 224 pp., illus., index.
(heavy on style and interiors)

Hamlin, Talbot, *Greek Revival Architecture in America.* Reprint ed. New York: Dover, 1969. 439 pp., illus., biblio.
(a classic reference on one of America's most enduring styles)

Ingle, Marjorie. *The Mayan Revival Style.* Salt Lake City: Peregrine Smith, 1984. 92 pp., illus., biblio., index.
(describes an esoteric and uniquely American mode)

Jordy, William H. *American Buildings and Their Architects*, Vol. 3, *Progressive and Academic Ideals at the Turn of the Twentieth Century.* Garden City, N.Y.: Doubleday, 1972. 448 pp., illus., gloss., index.

———. *American Buildings and Their Architects*, Vol. 4, *The Impact of European Modernism in the Mid-Twentieth Century.* Garden City, N.Y.: Doubleday, 1972. 496 pp., illus., gloss., index.

Kauffman, Henry J. *The American Farmhouse.* New York: Bonanza, 1979. 265 pp., illus., biblio., index.
(a good survey of early farmhouses in New England, the Middle States, and the South)

Kidney, Walter C. *The Architecture of Choice: Eclecticism in America, 1880–1930.* New York: Braziller, 1974. 178 pp., illus., notes, index.
(an excellent and well-illustrated work that documents a significant stylistic trend by lesser known architects)

Lancaster, Clay. *The American Bungalow, 1880s–1920s.* New York: Abbeyville Press, 1985. 256 pp., illus.
(an excellent reference on a popular American prototype)

———. *The Japanese Influence in America.* New York: Rawls/Twane, 1963. 292 pp., illus.
(a well-illustrated and well-documented book on an esoteric aspect of American architecture)

Loth, Calder and Julius T. Sadler, Jr. *The Only Proper Style: Gothic Architecture in America.* Boston: New York Graphic Society, 1976. 184 pp., illus., biblio.

Maddex, Diane, ed. *Master Builders: A Guide to Famous American Architects.* Washington, D.C.: Preservation Press, 1985. 192 pp., illus., biblio., index.
(an excellent, concise, portable guide examining the works of 40 major architects)

Makinson, Randell L. *Greene and Greene: Architecture as a Fine Art.* Salt Lake City: Peregrine Smith, 1979. 288 pp., illus.
(the classic reference on Japanese-influenced California redwood Arts and Crafts style residences)

McArdle, Alma deC. and Deirdre B. McArdle. *Carpenter Gothic: Nineteenth-Century Ornamented Houses of New England.* New York: Whitney Library of Design, 1978. 160 pp., illus., biblio., index.

Moore, Charles W., Kathryn Smith, and Peter Becker. *Home Sweet Home: American Domestic Vernacular Architecture.* New York: Rizzoli, 1983. 150 pp., illus.

Ochsner, Jeffrey Karl. *H.H. Richardson: Complete Architectural Works.* Cambridge: MIT Press, 1982. 466 pp., illus., index.
(an account of one of the titans of late nineteenth-century American architecture)

Pierson, William H., Jr. *American Buildings and Their Architects,* Vol. 1, *The Colonial and Neoclassical Styles.* Garden City, N.Y.: Doubleday, 1970. 503 pp., illus., gloss., index.
(a very thorough reference work)

———. *American Buildings and Their Architects*, Vol. 2A, *Technology and the Picturesque. The Corporate and the Early Gothic Styles.* Garden City, N.Y.: Doubleday, 1978. 500 pp., illus., gloss., index.

Robinson, Cervin and Rosemarie Haag Bletter. *Skyscraper Style: Art Deco New York.* New York: Oxford University Press, 1975. 198 pp., illus., notes.

Scully, Vincent, Jr. *The Shingle Style and the Stick Style: Architectural Theory and Design from Richardson to the Origins of Wright.* Rev. ed. New Haven: Yale University Press, 1971. 184 pp., illus., biblio., index.
(the classic book on this style)

———. *The Shingle Style Today,* or *The Historian's Revenge.* New York: Braziller, 1974. 118 pp., illus., index.

Shopsin, William C. and Mosette Glaser Broderick. Municipal Art Society. *The Villard Houses: Life Story of a Landmark.* New York: Viking, 1980. 144 pp., illus.
(discusses the origins of "group living" for the well-to-do; tells the story of those who flaunted their new-found wealth and social status at the end of the nineteenth century)

Spencer, Brian A., ed. *The Prairie School Tradition: Sullivan, Adler, Wright and Their Heirs.* New York: Whitney Library of Design, 1979. 304 pp., illus., biblio., index.
(the origins of modern architecture in America)

The Origins of Cast Iron Architecture in America. Intro. by W. Knight Sturges. Reprint, New York: Da Capo, 1970. (facsimile edition includes *Cast Iron Buildings: Their Construction and Advantages*, 1856, and *Illustrations of Iron Architecture*, 1865)

Ware, William R. *The American Vignola: A Guide to the Making of Classical Architecture.* Reprint ed. New York: W.W. Norton, 1977. 145 pp.
(a compact guide to the classical elements, proportions, and detailing)

Weitze, Karen. *California's Mission Revival: Transition to the Twentieth Century.* Los Angeles: Hennessey and Ingalls, 1984. 160 pp., illus.
(an important book on a dominant California Architectural tradition)

Whitehill, Muir and Frederic D. Nichols. *Palladio in America.* New York: Rizzoli, 1976. 120 pp., illus.
(a fascinating account of classical influences on eighteenth century American architecture)

Wilson, Richard Guy, Diane Pilgrim, and Richard N. Murray. *The American Renaissance:* 1876–1917. New York: Brooklyn Museum, 1979. 232 pp., illus.
(a well-illustrated account of the golden age of American monumental architecture)

PRESERVATION MOVEMENT

Fitch, James Marston. *Historic Preservation: Curatorial Management of the Built World.* New York: McGraw-Hill, 1982. 434 pp., illus., index.
(an important, scholarly work on preservation theory and historic interpretation)

Hosmer, Charles B., Jr. *Preservation Comes of Age: From Williamsburg to the National Trust, 1926–1949.* Charlottesville:University Press of Virginia, 1981. 2 vols., 1,291 pp., biblio., index.

National Trust for Historic Preservation. *All About Old Buildings: The Whole Preservation Catalog.* Diane Maddex, ed. Washington, D.C.:Preservation Press, 1985. 436 pp., illus., biblio., index.
(a fantastic compendium of sources and contacts for the layman as well as the professional)

National Trust for Historic Preservation, Tony P. Wren and Elizabeth D. Mulloy. *America's Forgotten Architecture.* New York: Pantheon, 1976. 312 pp., illus., biblio., index.

Thurber, Pamela, ed. *Controversies in Preservation: Understanding the Movement Today*, Preservation Policy Research Series. Washington, D.C.: National Trust for Historic Preservation, 1985. 60 pp.

Weinberg, Nathan C. *Preservation in American Towns and Cities.* Boulder, Colo.: Westview Press, 1979. 233 pp., illus., index.

Where to Look: A Guide to Preservation Information. Washington, D.C.: Advisory Council on Historic Preservation, 1982.

Williams, Norman, Jr., Edmund Kellogg, and Frank Gilbert, eds. *Readings in Historic Preservation: Why? What? How?* New Brunswick, N.J.: Center for Urban Policy Research, Rutgers University, 1983.

PRESERVATION GUIDELINES AND HANDBOOKS

A Primer: Preservation for the Property Owner. Albany: Preservation League of New York State, 1978. 40 pp., illus., biblio.
(contains practical, concise advice; the Preservation League, listed in the resource section, publishes other useful literature from time to time)

Bullock, Orin M. *The Restoration Manual: An Illustrated Guide to the Preservation and Restoration of Old Buildings.* Reprint. New York:Van Nostrand Reinhold, 1983. 192 pp., illus., gloss., biblio.
(focuses on museum-type preservation)

Glenn, Marsha. *Historic Preservation: A Handbook for Architecture Students.* Washington, D.C.: American Institute of Architects, 1974.

Hanson, Shirley and Nancy Hubby. *Preserving and Maintaining the Older Home.* New York: McGraw-Hill, 1983. 237 pp., illus., biblio., index.
(a very practical, well-illustrated guide suitable for the nonprofessional)

Hutchins, Nigel. *Restoring Old Houses of Brick and Stone.* New York: Van Nostrand Reinhold, 1983. 192 pp.

Insall, Donald. *The Care of Old Buildings Today: A Practical Guide.* London: Architectural Press, 1972. 197 pp., illus., biblio.
(technically excellent, but quintessentially British; terminology and some techniques not applicable to American buildings)

Keune, Russell V., ed. *The Historic Preservation Yearbook.* Bethesda, Md.: Adler and Adler, 1984. 590 pp.

Labine, Clem and Carolyn Flaherty, eds. *The Original Old-House Journal Compendium.* Woodstock, N.Y.: Overlook Press, 1980. 400 pp., illus., index.
(an excellent, continuing series of helpful advice and techniques, with lists of sources and suppliers)

National Trust for Historic Preservation, Information Series. *Basic Preservation Procedures.* Rev. ed. Washington, D.C.: Preservation Press, 1983. 20 pp., biblio.

North American International Regional Conference, Williamsburg, Va., and Philadelphia, 1972. *Preservation and Conservation: Principles and Practices.* Washington, D.C.: Preservation Press, 1976.

Poore, Patricia and Clem Labine, eds. *The Old-House Journal New Compendium: A Complete How-to Guide for Sensitive Rehabilitation.* New York: Doubleday, 1983. 426 pp., illus., gloss., index.

Stahl, Frederick A. *A Guide to the Maintenance, Repair, and Alteration of Historic Buildings.* New York: Van Nostrand Reinhold, 1984. 224 pp., illus., index.

Technical Preservation Services, U.S. Department of the Interior. *Respectful Rehabilitation: Answers to Your Questions About Old Buildings.* Washington, D.C.: Preservation Press, 1982. 192 pp., illus., biblio., index.
(a popular, illustrated "Dear Abby" format; discusses what to do and what not to do with old buildings)

The Secretary of the Interior's Standards for Rehabilitation and Guidelines for Rehabilitating Historic Buildings. Rev. ed. Washington, D.C.: Technical Preservation Services, U.S. Department of the Interior, 1983. 61 pp.
(the "Ten Commandments" of preservation; a very strict interpretation of dos and don'ts)

PRESERVATION: LEGAL, ECONOMIC, AND SOCIAL ISSUES

Appraising Easements: Guidelines for Valuation of Historic Preservation and Land Conservation Easements. National Trust for Historic Preservation and Land Trust Exchange. Washington, D.C.: National Trust, 1984. 68 pp., biblio.

Carrington, Merrill Ware. *Design Guidelines: An Annotated Bibliography.* Washington, D.C.: National Endowment for the Arts, 1977. 30 pp.

Chittenden, Betsy, with Jacques Gordon. *Older and Historic Buildings and the Preservation Industry.* National Trust for Historic Preservation, Preservation Policy Research Series, Washington, D.C.: 1984. 19 pp., biblio.

Conservation Foundation, National Trust for Historic Preservation, and American Bar Association. *Reusing Old Buildings: Preservation Law and the Development Process.* Washington, D.C.: National Trust, 1984. 450 pp.

Costonis, John. Space Adrift: *Saving Urban Landmarks Through the Chicago Plan.* Champaign: University of Illinois Press, 1974. 228 pp., illus., biblio., index.
(a very specialized theoretical argument)

Duerksen, Christopher J., ed. *A Handbook on Preservation Law.* Washington, D.C.: Conservation Foundation, 1983. 523 pp., biblio., index.

Federal Tax Incentives for the Rehabilitation of Historic Buildings. National Trust for Historic Preservation, Information Series. Washington, D.C.: Preservation Press, 1984. 7 pp., illus.

Fisher, Charles E., William G. MacRostie, and Christopher A. Sowick. *Director of Historic Preservation Easement Organizations.* Technical Preservation Services, U.S. Department of the Interior. Washington, D.C.: 1981. 23 pp.

Listokin, David. *Landmarks Preservation and the Property Tax: Assessing Landmark Buildings for Real Taxation Purposes.* New Brunswick, N.J.: Center for Urban Policy Research, 1982. 248 pp., illus., biblio.

National Trust for Historic Preservation. *Economic Benefits of Preserving Old Buildings.* Washington, D.C.: Preservation Press, 1976. 168 pp., illus.
(good illustrative examples, but somewhat dated)

National Trust for Historic Preservation. Richard Wright, consultant. *A Guide to Delineating Edges of Historic Districts.* Washington, D.C.:Preservation Press, 1976. 96 pp., illus., gloss., biblio.

Parrott, Charles. *Access to Historic Buildings for the Disabled: Suggestions for Planning and Implementation.* Technical Preservation Services. U.S. Department of the Interior, Technical Report. Washington, D.C.: 1980. 92 pp., illus., biblio., (NTIS no. PB85-180826)

Paseltiner, Ellen Kettler and Deborah Tyler. *Zoning and Historic Preservation: A Survey of Current Zoning Techniques in U.S. Cities to Encourage Historic Preservation.* Rev. ed. Chicago: Landmarks Preservation Council of Illinois, (407 South Dearborn Street, 60605) 1984. 33 pp.

Practising Law Institute Handbooks: Historic Preservation Law (1981, 1982). *Rehabilitating Historic Buildings* (1983). *Rehabilitating Historic Properties* (1984). New York: Practising Law Institute (810 Seventh Avenue, 10019).

Reynolds, Judith. *Historic Properties: Preservation and the Valuation Process.* Chicago: American Institute of Real Estate Appraisers (430 North Michigan Avenue), 1982. 115 pp., illus., biblio., index.

Roddewig, Richard J. *Preparing a Historic Preservation Ordinance,* PAS Report 374. Chicago: American Planning Association, 1983. 46 pp., illus.

Wolf, Peter. *Land in America: Its Value, Use and Control.* New York: Pantheon, 1981. 591 pp., illus., biblio., index. (not limited to preservation)

Warner, Raynor M., Sibyl Groff, and Ranne P. Warner. *Business and Preservation: A Survey of Business Conservation of Buildings and Neighborhoods.* New York: INFORM, 1978. 295 pp., illus. (recycling as a business investment; includes very interesting case studies)

Zick, Steven J. *Preservation Easements: The Legislative Framework,* Preservation Policy Research Series, Washington, D.C.: National Trust for Historic Preservation, 1984. 51 pp.

HISTORICAL DOCUMENTATION AND RESEARCH

Borchers, Perry E. *Photogrammetric Recording of Cultural Resources.* Technical Preservation Services, U.S. Department of the Interior, Technical Report. Washington, D.C.: 1977. 38 pp., illus., (NTIS no. PB85-180792).

Carter, Margaret. *Researching Heritage Buildings.* Ottawa: Ministry of the Environment, Parks Canada, 1983. 38 pp., illus. (very helpful booklet on research methods and sources of information)

Chambers, Henry J. *Rectified Photography and Photo Drawings for Historic Preservation.* Technical Preservation Services, U.S. Department of the Interior, Technical Report. Washington, D.C.: 1973. 37 pp., illus., (NTIS no. PB85-180768).

Chitham, Robert. *Measured Drawings for Architects.* New York: Nichols Publishing, 1980. 128 pp., illus., biblio., index.

Dean, Jeff. *Architectural Photography: Techniques for Architects, Preservationists, Historians, Photographers, and Urban Planners.* American Association for State and Local History. Nashville: 1982. 144 pp., illus., biblio., index.

Ellsworth, Linda. *The History of a House: How to Trace It.* American Association for State and Local History, Technical Leaflet 89. Nashville: 1976. 8 pp., biblio.

Fire Insurance Maps in the Library of Congress. Comprehensive listing by staff, Geography and Map Division. Washington, D.C.: Library of Congress, 1981. 783 pp., illus., index (GPO no. 030-004-00018-3).

Gebhard, David and Deborah Nevins. *Two-Hundred Years of American Architectural Drawing.* New York: Whitney Library of Design, 1977. 306 pp., illus., biblio., index.

Hart, David M. *X-Ray Examination of Historic Structures.* Technical Preservation Services, U.S. Department of the Interior, Technical Report. Washington, D.C.: 1975. 24 pp., illus., (NTIS no. PB85-180800).

Historic America: Buildings, Structures, and Sites Recorded by the Historic American Buildings Survey and the Historic American Engineering Record. Comprehensive listing by archivist Alicia Stamm, essays edited by curator C. Ford Peatross. Washington, D.C.: Library of Congress, 1983. 724 pp., illus., (GPO no. 030-000-00149-4).

Kalman, Harold. *The Evaluation of Historic Buildings.* Ottawa: Parks Canada, 1980. 39 pp., illus.
(excellent methodology for what to include in a survey)

Lugo, Lelahvon. *Library Resources in Washington, D.C., Relating to Historic Preservation.* Washington, D.C.: National Trust for Historic Preservation, 1977. 55 pp.

Massey, James C. *How to Organize an Architectural Survey.* Washington, D.C.: National Trust for Historic Preservation, 1976.

McKee, Harley J. *Recording Historic Buildings.* Historic American Buildings Survey, U.S. Department of the Interior. Washington, D.C.: 1976. 176 pp., illus., biblio., index.
(the classic work on this subject; describes various techniques for field notes, measured drawings, and photographic documentation)

McQuaid, James, ed. *An Index to American Photographic Collections.* Boston: G.K. Hall, 1982. 40 pp., illus., index.

Panoramic Maps of Cities in the United States and Canada: A Checklist of Maps in the Collections of the Library of Congress, Geography and Map Division. Compiled by John R. Hébert, revised by Patrick E. Dempsey. 2nd ed. Washington, D.C.: Library of Congress, 1984. 189 pp., illus., index, (GPO no. 030-004-00022-1).

Shulman Julius. *The Photography of Architecture and Design: Photographing Buildings, Interiors and the Visual Arts.* New York: Whitney Library of Design, 1977. 238 pp., illus., index.

Travers, Jean and Susan Shearer. *Guide to Resources Used in Historic Preservation Research.* Washington, D.C.: National Trust for Historic Preservation, 1978. 26 pp.

Wilson, Rex. *Archeology and Preservation.* National Trust for Historic Preservation, Information Series. Washington, D.C.: Preservation Press, 1980. 20 pp., biblio.

REHABILITATION AND ADAPTATION: RURAL AND URBAN PROBLEMS

Adaptive-Use Development: Economics, Process, and Profiles. Washington, D.C.: Urban Land Institute, 1978. 246 pp., illus., biblio.
(provides specific cost data on diverse adaptive-use projects of varying sizes and types, including commercial projects)

Brolin, Brent. *Architecture in Context: Fitting New Buildings with Old.* New York: Van Nostrand Reinhold, 1979. 144 pp., illus.

Bunnell, Gene. *Built to Last: A Handbook on Recycling Old Buildings.* Washington, D.C.: Preservation Press, 1977. 126 pp., illus., index.

Cantacuzino, Sherban, ed. *Architectural Conservation in Europe.* New York: Whitney Library of Design, 1975. 144 pp., illus., index.
(European examples of adaptive use, including many different building types)

———. *New Uses for Old Buildings.* New York: Whitney Library of Design, 1975. 264 pp., illus.

———. And Susan Brandt, *Saving Old Buildings.* New York: Nichols Publishing, 1981. 230 pp., illus.

Cawley, Frederick D., ed. *Historic Landscape Preservation and Restoration: An Annotated Bibliography for New York State.* Preservation League of New York State. Albany: 1977. 6 pp.

Closs, Christopher W. *Preserving Large Estates.* National Trust for Historic Preservation, Information Series. Washington, D.C.: Preservation Press, 1982. 24 pp., illus., biblio.

Curtis, John O. *Moving Historic Buildings.* Technical Preservation Services, U.S. Department of the Interior, Technical Report. Washington, D.C.: 1979. 56 pp., illus., biblio., (GPO no. 024-005-00857-8).

Diamonstein, Barbaralee. *Buildings Reborn: New Uses, Old Places.* New York: Harper & Row, 1978. 255 pp., illus., biblio., index.

Ewald, William, R., Jr., and Daniel R. Mandelker. *Street Graphics: A Concept and a System.* 2nd ed. Washington, D.C.: American Society of Landscape Architects Foundation, 1977. 176 pp., illus., biblio.
(sign ordinances, graphics, architectural lettering, and roadscapes)

Favretti, Rudy J. and Joy Putman Favretti. *Landscapes and Gardens for Historic Buildings: A Handbook for Reproducing and Creating Authentic Landscape Settings.* Nashville: American Association for State and Local History Press, 1979. 202 pp., illus.

Finegold, Anderson Notter. *Recycling Historic Railroad Stations: A Citizen's Manual.* U.S. Department of Transportation, Office of the Secretary, Room 10223. Washington, D.C.: 1978. 83 pp., illus.

Fleming, Ronald Lee. *Facade Stories: Changing Faces of Main Street Storefronts and How to Care for Them.* New York: Hastings House, 1982. 128 pp., illus., biblio., gloss., index.

Hawley, Peter and National Main Street Center. *The Main Street Book: A Guide to Downtown Revitalization.* Washington, D.C.: Preservation Press, 1986. 324 pp., illus., biblio., index.

Holdsworth, Deryck, ed., Heritage Canada Foundation. *Reviving Main Street.* Buffalo: University of Toronto Press, 1985. 256 pp.

Jandl, H. Ward. *Rehabilitating Historic Storefronts.* Technical Preservation Services, U.S. Department of the Interior, Preservation Brief No. 11. Washington, D.C.: 1982. 12 pp., illus., (GPO no. 024-005-00886-1).

McNulty, Robert H. and Stephen A. Kliment, eds. *Neighborhood Conservation: A Handbook of Methods and Techniques.* Reprint. New York: Whitney Library of Design, 1979. 288 pp., illus., index.

Mintz, Norman. *A Practical Guide to Storefront Rehabilitation.* Preservation League of New York State, Technical Series, No. 2. Albany:1982. 8 pp., illus.

Movie Palaces: Renaissance and Reuse. Educational Facilities Laboratories, 680 Fifth Avenue, New York: 1982. 120 pp., illus., biblio.

Natonal Trust for Historic Preservation. *Old and New Architecture: Design Relationship.* Washington, D.C.: Preservation Press, 1980. 280 pp., illus., biblio., index.
(a very good series of illustrated articles on the problems of inserting new buildings in historic areas)

Ramati, Raquel. *How to Save Your Own Street.* Urban Design Group, New York Department of City Planning, Garden City, N.Y.: Dolphin Books, Doubleday, 1981. 159 pp., illus.

Reed, Richard Ernie. *Return to the City: How to Restore Old Buildings and Ourselves in America's Historic Urban Neighborhoods.* Garden City, N.Y.: Doubleday, Dolphin, 1979. 191 pp., illus.

Rifkind, Carole. *Main Street: The Face of Urban America.* New York: Harper & Row, 1977. 267 pp., illus., biblio., index.

Rudofsky, Bernard. *Streets for People: A Primer for Americans.* 2nd ed. New York: Van Nostrand Reinhold, 1982. 351 pp., illus.

Shopsin, William C. and Grania Bolton Marcus. *Saving Large Estates: Conservation, Historic Preservation, Adaptive Reuse.* Setauket, N.Y.: Society for the Preservation of Long Island Antiquities, 1977. 199 pp., illus., biblio.

Stratton, Jim. *Pioneering in the Urban Wilderness: All About Lofts.* New York: Urizen Books, 1977. 208 pp.
(documents loft living, a recent American trend)

Ziegler, Arthur P., Jr. *Historic Preservation in Inner City Areas: A Manual of Practice.* Rev. ed. Pittsburgh: Ober Park Associates, 450 Landmarks Building. 1974. 85 pp., illus.

REHABILITATION, REPLACEMENT, AND MAINTENANCE OF SPECIFIC MATERIALS OR PARTS

Association for Preservation Technology. *The Victorian Design Book: A Complete Guide to Victorian House Trim.* Ottawa: Firefly Books, 1984. 416 pp.

Batcheler, Penelope Hartshorne. *Paint Color Research and Restoration.* American Association for State and Local History, Technical Leaflet 15. Nashville: 1968. 4 pp.

Berryman, Nancy D. and Susan M. Tindall. *Terra Cotta: Preservation of a Historic Building Material.* Chicago: Landmarks Preservation Council of Illinois (407 South Dearborn Street, 60605), 1983. 38 pp., illus.

Brolin, Brent C. and Jean Richards. *Sourcebook of Architectural Ornaments: Designers, Craftsmen, Manufacturers, and Distributors of Exterior Architectural Ornament.* New York: Van Nostrand Reinhold, 1982. 288 pp., illus., index.

Cawley, Frederick D. *Property Owner's Guide to Paint Restoration and Preservation.* Preservation League of New York State, Technical Series, No. 1. Albany: 1976. 8 pp., biblio.

Chambers, Henry J. *Cyclical Maintenance for Historic Buildings.* Technical Preservation Services, U.S. Department of the Interior, Technical Report. Washington, D.C.: 1976. 125 pp., illus., biblio., (GPO no. 024-005-00637-1).

Exterior Decoration: Victorian Colors for Victorian Houses. Philadelphia: The Athenaeum of Philadelphia, 1975. 95 pp., biblio.
(a facsimile edition of Devoe Paint Catalogue of 1885 with sample color chips and suggestions for alternate color schemes; illustrated with engravings of period buildings)

Gayle, Margot and David W. Look. *Metals in America's Historic Buildings: Uses and Preservation Treatments.* Technical Preservation Services, U.S. Department of the Interior, Technical Report. Washington, D.C.: 1980. 168 pp., illus., (GPO no. 024-005-00910-8).
(excellent technical reference work)

Goodall, Harrison and Renee Friedman. *Log Structures: Preservation and Problem Solving.* Nashville: American Association for State and Local History, 1980. 119 pp., illus., biblio., index.

Grimmer, Anne E. *A Glossary of Historic Masonry Deterioration Problems and Preservation Treatments.* Technical Preservation Services, U.S. Department of the Interior, Technical Report. Washington, D.C.: 1984. 68 pp., illus., (GPO no. 024-005-00870-5).

———. *Dangers of Abrasive Cleaning to Historic Buildings.* Technical Preservation Services, U.S. Department of the Interior, Preservation Brief No. 6. Washington D.C.: 1979. 8 pp., illus., (GPO no. 024-005-00882-9).

Hutchins, Nigel. *Restoring Houses of Brick and Stone.* New York: Van Nostrand Reinhold, 1983. 192 pp.

Litchfield, Michael and Rosemarie Hausherr. *Salvaged Treasures: Designing and Building with Architectural Salvage.* New York: Van Nostrand Reinhold, 1983. 253 pp., illus., append., biblio., index.
(very well illustrated, practical suggestions on incorporating old building elements).

Mack, Robert C. *Repainting Mortar Joints in Historic Brick Buildings.* Technical Preservation Services, U.S. Department of the Interior, Preservation Brief No. 2. Washington, D.C.: 1980. 8 pp., illus., (GPO no. 024-005-00878-1).

———. *The Cleaning and Waterproof Coating of Masonry Buildings.* Technical Preservation Services, U.S. Department of the Interior, Preservation Brief No. 1. Washington, D.C.: 1975. 4 pp., illus., (GPO no. 024-005-00877-2).

McKee, Harley J. *Introduction to Early American Masonry: Stone, Brick, Mortar and Plaster.* Washington, D.C.: Preservation Press, 1973. 92 pp., illus., biblio., index.
(the classic reference on this subject; very instructive)

Moss, Roger. *Century of Color: Exterior Decoration for American Buildings, 1820–1920.* Watkins Glen, N.Y.: American Life Foundation, 1981. 112 pp., illus., gloss., index.

Myers, John H. *The Repair of Historic Wooden Windows.* Technical Preservation Services, U.S. Department of the Interior, Preservation Brief No. 9. Washington, D.C.: 1981. 8 pp., illus., (GPO no. 024-005-00885-3).

——— and Gary L. Hume. *Aluminum and Vinyl siding on Historic Buildings: The Appropriateness of Substitute Materials for Resurfacing Historic Wood Frame Buildings.* Technical Preservation Services, U.S. Department of the Interior. Preservation Brief No. 8. Rev. ed. Washington, D.C.: 1984. 8 pp., illus., (GPO no. 024-005-00869-1).

National Slate Association, *Slate Roofs.* Rev. ed. Fairhaven, Vermont: Vermont Structural Slate Company, 1977.

Phillips, Morgan W. and Judith E. Selwyn. *Epoxies for Wood Repairs in Historic Buildings.* Technical Preservation Services, U.S. Department of the Interior, Technical Report. Washington, D.C.: 1978. 72 pp., illus., (NTIS no. PB85-180834).

Smith, Baird M. *Moisture Problems in Historic Masonry Walls: Diagnosis and Treatment.* Technical Preservation Services, U.S. Department of the Interior, Technical Report. Washington, D.C.: 1984. 48 pp., illus., (GPO no. 024-005-00872-1).

Southworth, Susan and Michael. *Ornamental Ironwork: An Illustrated Guide to Its Design, History, and Use in American Architecture.* Boston: David R. Godine, 1978. 202 pp., illus., biblio., index.

Sweetser, Sarah M. *Roofing for Historic Buildings.* Technical Preservation Services, U.S. Department of the Interior, Preservation Brief No. 4. Washington, D.C.: 1978. 8 pp., illus., (GPO no. 024-005-00880-2).

The Old-House Journal 1985 Catalog: A Buyer's Guide for the Pre-1939 House. Brooklyn, New York: Old-House Journal, 1984. 212 pp., illus., index.

The Preservation of Historic Adobe Buildings. Technical Preservation Services, U.S. Department of the Interior, Preservation Brief No. 5. Washington, D.C.: 1978. 8 pp., illus., (GPO no. 024-005-00881-1).

The Preservation of Historic Pigmented Structural Glass (Vitrolite and Carrara Glass). Technical Preservation Services, U.S. Department of the Interior, Preservation Brief No. 12. Washington, D.C.: 1984. 8 pp., illus., (GPO no. 024-005-00851-9).

Tiller, de Teel Patterson. *The Preservation of Historic Glazed Architectural Terra-Cotta.* Technical Preservation Services, U.S. Department of the Interior, Preservation Brief No. 7. Washington, D.C.:1979. 8 pp., illus., (GPO no. 024-005-0883-7).

Waite, Diana S. *Architectural Elements: The Technological Revolution.* Princeton, N.J.: The Pyne Press/American Catalog Collection, 1972. 70 pp., illus.
(reproduction of nineteenth-century manufacturers' catalogs, includes galvanized iron roof plates and corrugated sheets; cast iron facades, columns, door and window caps, sills, and lintels; galvanized cornices; marbleized slate mantels; plumbing and heating supplies, and fixtures; staircases, balconies, newels, and balusters in wood and iron; cut and etched glass transoms and side lights)

Weeks, Kay D. and David W. Look. *Exterior Paint Problems on Historic Woodwork.* Technical Preservation Services, U.S. Department of the Interior, Preservation Brief No. 10. Washington, D.C.: 1982. 12 pp., illus., (GPO no. 024-005-00885-3).

Weiss, Norman R. *Exterior Cleaning of Historic Masonry Buildings.* Technical Preservation Services, U.S. Department of the Interior, Technical Report. Washington, D.C.: 1977. 18 pp., biblio, (NTIS no. PB85-180818).

REHABILITATION: MECHANICAL SYSTEMS AND ENERGY CONSERVATION

Myers, Denys Peter. *Gaslighting in America: A Guide to Historic Preservation.* Technical Preservation Services, U.S. Department of the Interior, Technical Report. Washington, D.C.: 1978. 248 pp., illus. biblio, index, (NTIS no. PB85-192201)

Park Sharon C. *The Repair and Thermal Upgrading of Historic Steel Windows.* Technical Preservation Services, U.S. Department of the Interior, Preservation Brief No. 13. Washington, D.C.: 1984. 12 pp., illus. (GPO no. 024-005-00868-3).

———. *Improving Thermal Efficiency: Historic Wooden Windows.* The Colcord Building, Oklahoma City, Oklahoma. Technical Preservation Services, U.S. Department of the Interior, Preservation Case Study, Washington, D.C.: 1982. 16 pp., illus., (GPO no. 024-005-00840-3).

Smith, Baird M. *Conserving Energy in Historic Buildings.* Technical Preservation Services, U.S. Department of the Interior, Preservation Brief No. 3. Washington, D.C.: 1978. 8 pp., illus., (GPO no. 024-005-00879-9).

INTERIORS: HISTORICAL BACKGROUND OF DECORATIVE ARTS

Andersen, Timothy J., Eudorah M. Moore, and Robert W. Winter, eds. *California Design 1910.* Salt Lake City: Peregrine Smith, 1980. 144 pp., illus., index.
(chronicles the Arts and Crafts Movement on the West Coast)

Bishop, Robert and Patricia Coblentz. *American Decorative Arts: 360 Years of Creative Design.* New York: Abrams, 1982. 406 pp., illus., biblio., index.

Clark, Robert Judson. *The Arts and Crafts Movement in America 1876–1916.* Princeton: Princeton University Press, 1972. 190 pp., illus., biblio.
(a very useful survey relating to decorative arts and furniture)

Frangiamore, Catherine Lynn. *Wallpapers in Historic Preservation.* Technical Preservation Services, U.S. Department of the Interior, Technical Report. Washington, D.C.: 1977. 56 pp., illus., (NTIS no. PB85-180784).

Kouwenhoven, John A. *The Columbia Historical Portrait of New York: An Essay in Graphic History in Honor of the Tricentennial of New York City and the Bicentennial of Columbia University.* Reprint. New York: Octagon Books, 1983. illus.

Lambourne, Lionel. *Utopian Craftsmen: The Arts and Crafts Movement from the Cotswolds to Chicago.* New York: Van Nostrand Reinhold, 1982. 240 pp., illus.

Ludwig, Coy L. *The Arts and Crafts Movement in New York State, 1890s–1920s.* Salt Lake City: Peregrine Smith, 1984. 128 pp., illus.

Mumford, Lewis. *The Brown Decades: A Study of the Arts in America, 1865–1895.* New York: Dover, 1955. 266 pp., illus., biblio.
(a classic on Victorian deign and taste)

Nylander, Jane C. *Fabrics for Historic Buildings.* 3rd ed. Washington, D.C.: Preservation Press, 1983. 160 pp., illus., gloss., biblio.

Oman, Charles C. and Jean Hamilton. *Wallpapers: An International History and Illustrated Survey from the Victoria and Albert Museum.* New York: Abrams, 1982. 464 pp., illus., biblio., index.

Picture Book of Authentic Mid-Victorian Gas Lighting Fixtures. Reprint of Mitchell, Vance and Company catalog with new introduction by Denys Peter Myers. New York: Dover, 1985.

Rogers, Meyrick R. *American Interior Design: The Traditions and Development of Domestic Design from Colonial Times to the Present.* Reprint. New York: Ayer, 1976. illus.
(an excellent illustrated history of American interiors; describes architectural detals and furnishings)

Sanders, Barry. *Gustav Stickley: The Craftsman Movement in America.* Salt Lake City: Peregrine Smith, 1979. 240 pp., illus.

Seale, William. *The Tasteful Interlude: American Interiors Through the Camera's Eye, 1860–1917.* Rev. ed. Nashville: American Association for State and Local History Press, 1980. 288 pp., illus., index.
(contains very good period illustrations of various types of interiors from the fashionable drawing room to a Yukon cabin)

———. *Recreating the Historic House Interior.* Nashville: American Association for State and Local History Press, 1979. 270 pp., illus., biblio., index.
(a very helpful guide to the practical problems of historic house furnishings)

Stickley, Gustav. *Craftsman Homes: Architecture and Furnishings of the American Arts and Crafts Movement.* Reprint ed. New York: Dover, 1979. 224 pp., illus.

NOTE: Publications issued by the Technical Preservation Services may be ordered by GPO number from the Superintendent of Documents, U.S. Government Printing Office, Washington, D.C. 20402 or by NTIS number from the U.S. Department of Commerce, National Technical Information Service, 5285 Port Royal Road, Springfield, Va. 22161.

PERIODICALS

Alert
Preservation Action
1700 Connecticut Avenue, N.W.
Washington, D.C. 20009
(keeps abreast of the latest federal legislation)

APT Bulletin
Association for Preservation Technology
P.O. Box 2487, Station D
Ottawa, Ontario K1P 5W6, Canada
(a highly technical professional journal)

Architectural Record
McGraw-Hill Publications Company
1221 Avenue of the Americas
New York, N.Y. 10020

Architecture: The AIA Journal and local chapter magazines
American Institute of Architects
1735 New York Avenue, N.W.
Washington, D.C. 20006

Blueprints
National Building Museum
440 G Street, N.W.
Washington, D.C. 20001

Canadian Heritage
The Heritage Canada Foundation
P.O. Box 1358, Station B
Ottawa, Ontario K1P 5R4, Canada

Environmental Comment
Urban Land Institute
1090 Vermont Avenue, N.W.
Washington, D.C. 20005

Historic Preservation,
Preservation News
(includes "Marketplace" section) and
Preservation Law Reporter
National Trust for Historic Preservation
1785 Massachusetts Avenue, N.W.
Washington, D.C. 20036

History News
American Association for State and Local History
708 Berry Road
Nashville, Tenn. 37204

Landscape Architecture
American Society of Landscape Architects
1733 Connecticut Avenue, N.W.
Washington, D.C. 20009

Livability Digest
Partners for Livable Places
1429 21st Street, N.W.
Washington, D.C. 20036

Metropolis
207 West 25th Street
New York, N.Y. 10017

Museum News
American Assocation of Museums
1005 Thomas Jefferson Street, N.W.
Washington, D.C. 20007

Old-House Journal
(includes "The Old-House Emporium" section) and
The Old-House Journal Compendium
69-A Seventh Avenue
Brooklyn, N.Y. 11217

Planning
American Planning Assocation
1313 East 60th Street
Chicago, Ill. 60637

Progressive Architecture
600 Summer Street
Stamford, Conn. 06904

SIA Newsletter and *IA Journal*
Society for Industrial Archeology
National Museum of American History
Room 5020
Washington, D.C. 20560

Skyline
c/o Rizzoli Publications
597 Fifth Avenue
New York, N.Y. 10017

Society of Architectural Historians Journal, *SAH Forum*, and *Newsletter*
1700 Walnut Street
Philadelphia, Pa. 19103
(very academic and scholarly)

Technology and Conservation of Art, Architecture, and Antiquities
The Technology Organization Inc.
One Emerson Place
Boston, Mass. 02114
(contains excellent technical information for professionals and specialists in the preservation field)

The Victorian
The Victorian Society in America
219 South Sixth Street
Philadelphia, Pa. 19106
(contains popular articles on 19th-century architecture, furnishings, and taste)

RESOURCE ORGANIZATIONS AND AGENCIES

HISTORIC AND CIVIC ASSOCIATIONS

American Association for State and Local History (AASLA)
708 Berry Road
Nashville, Tenn. 37204
(issues very good publications on preservation subjects)

American Life Foundation
P.O. Box 349
Watkins Glen, N.Y. 14891
(a source of excellent reprints and facsimile editions of nineteenth-century books on taste, paint colors, furnishings, and so forth)

American Planning Association
1313 East 60th Street
Chicago, Ill. 60637

Association for Preservation Technology
P.O. Box 2487, Station D
Ottawa, Ontario K1P 5W6, Canada
(provides training courses and workshops for various facets of architectural conservation both in Canada and the United States)

Friends of Cast-Iron Architecture
235 East 87th Street
Room 6C
New York, N.Y. 10028
(a good source of technical information)

Friends of Terra Cotta
P.O. Box 421393
Main Post Office
San Francisco, Calif. 94142
(a good source of technical information, craftsmen, and supplies)

Heritage Canada Foundation
P.O. Box 1358, Station B
Ottawa, Ontario K1P 5 R4, Canada

National Trust for Historic Preservation
1785 Massachusetts Avenue, N.W.
Washington, D.C. 20036
(maintains historic properties as house museums; provides professional advice on all preservation issues; maintains a library and resource facilites; administers grants and loans; publishes a monthly newspaper *(Preservation News)*, a bimonthly magazine *(Historic Preservation)*, and various books and brochures *(Preservation Press)*; operates the Preservaton Shop, 1600 H Street, N.W., Washington, D.C. 20006, which sells books and other items.)

National Center for Preservation Law
2101 L Street, N.W.
Washington, D.C. 20037

Partners for Livable Places
1429 21st Street, N.W.
Washington, D.C. 20036
(places special emphasis on restoring public spaces and on tourism related to historic preservation)

Preservaton Action
1700 Connecticut Avenue, N.W.
Suite 401
Washington, D.C. 20009
(a lobbying group for favorable treatment of historic property in federal legislation and funding programs)

Preservation League of New York State
307 Hamilton Street
Albany, N.Y. 12210

Project for Public Spaces
875 Avenue of the Americas
Room 201
New York, N.Y. 10001
(provides an innovative analysis of public parks, plazas, and playgrounds)

Society for the Preservation of Long Island Antiquities
93 North County Road
Setauket, N.Y. 11733

Society for the Preservation of New England Antiquities
141 Cambridge Street
Boston, Mass. 02114

Urban Land Institute
1090 Vermont Avenue, N.W.
Suite 300
Washington, D.C. 20005

The Victorian Society in America
219 South Sixth Street
Philadelphia, Pa. 19106

GOVERNMENTAL AND ALLIED AGENCIES

Advisory Council on Historic Preservation
1100 Pennsylvania Avenue, N.W.
Suite 809
Washington, D.C. 20005

Archives of American Art
Smithsonian Institution
8th and G Streets, N.W.
Washington, D.C. 20560

Cooper-Hewitt Museum of Decorative Arts and Design
2 East 91st Street
New York, N.Y. 10028
(a great archive of the decorative arts; has displays, exhibits, a research collection, an excellent library; America's answer to the Victoria and Albert Museum)

Council on Environmental Quality
722 Jackson Place, N.W.
Washington, D.C. 20006

Forest Products Laboratory
U.S. Department of Agriculture
P.O. Box 5130
Madison, Wis. 53705

Institute for Applied Technology/Center for Building Technology
National Bureau of Standards
U.S. Department of Commerce
Washington, D.C. 20234

Library of Congress
First Street, N.E.
Washington, D.C. 20540

Prints and Photographs Division
(maintains the collection of photos, drawings, and documents compiled by the *HABS/HAER* program (original records and microfilm or microfiche copies are available to researchers); maintains other notable collections of prints, negatives, and transparencies documenting rural and urban architecture, 1800s to 1940s; serves as national center for *COPAR*, an agency founded to collect architectural records, and make the records available to researchers)

Geography and Map Division
(maintains collection of detailed fire-insurance maps (Sanborn Collection), railroad maps, and panoramic maps or illustrated atlases of nineteenth- and early twentieth-century cities and towns)

National Archives
Still Pictures Branch
Audiovisual Archives Division
8th Street and Pennsylvania Avenue, N.W.
Washington, D.C. 20408

National Alliance of Preservation Commissions
Hall of the States
444 North Capitol Street, N.W.
Suite 332
Washington, D.C. 20001

National Conference of State Historic Preservation Officers
Hall of the States
444 North Capitol Street, N.W.
Suite 332
Washington, D.C. 20001

National Conference of States on Building Codes and Standards
481 Carlisle Drive
Herndon, Va. 22101

National Park Service
U.S. Department of the Interior
Washington, D.C. 20013-7127

Archaeological Assistance Division
(helps to survey, excavate, and preserve archaeological sites)

Historic American Buildings Survey Historic American Engineering Record
(continues to survey and make detailed records of historic structures and sites)

National Register of Historic Places Interagency Resources Division
(oversees the *National Register of Historic Places*, the official list of properties classified as distinctive, historic districts, sites, buildings, structures, or objects; screens nominations for the listing; establishes preservation guidelines and approves state programs. *Federal Register* publishes cumulative list (includes over 37,500 items) every February)

National Historic Landmarks Program
History Division
(evaluates eligibility of properties on National Register that may also be designated as *National Historic Landmarks*)

Preservation Assistance Division
(collects information and issues publicatons on technical matters of repair and maintenance through the *Technical Preservation Services Branch*; supervises conformance with program guidelines for federal tax incentives for historic preservation)

State Historic Preservation Office
(Contact Department of Parks and Recreation, Department of Natural Resources, Office of Cultural Preservation, or State Historical Society within each state for exact location; except for offices in Anchorage, Alaska, and Vermillion, South Dakota, the location is in the capital city)

Western Archaeological Center
Divison of Adobe/Stone Conservation
National Park Service
U.S. Department of the Interior
P.O. Box 41058
Tucson, Ariz. 85717

PROFESSIONAL AND TRADE ORGANIZATIONS

American Concrete Institute
P.O. Box 19150
Detroit, Mich. 48219

American Institute of Architects
1735 New York Avenue, N.W.
Washington, D.C. 20006

American Society of Civil Engineers
345 East 47th Street
New York, N.Y. 10017

American Society of Interior Designers
Historic Preservation Committee
1430 Broadway
New York, N.Y. 10018

American Society of Landscape Architects
Historic Preservation Committee
1733 Connecticut Avenue, N.W.
Washington, D.C. 20009

American Wood Council
1619 Massachusetts Avenue, N.W.
Washington, D.C. 20036

Brick Institue of America
11490 Commerce Park Drive
Reston, Va. 22091

Building Stone Institute of America
420 Lexington Avenue
New York, NY 10017

Masonry Research Foundation
815 15th Street, N.W.
Washington, D.C. 20005

Society of Architectural Historians
1700 Walnut Street
Suite 716
Philadelphia, Pa. 19103

National Association of Historic Preservation Attorneys
P.O. Box 45, Century Station
Raleigh, N.C. 27602

INDEXES AND COLLECTIONS

American Institute of Architects Foundation
Prints and Drawings Collection
1735 New York Avenue, N.W.
Washington, D.C. 20006
(AIA also maintains a library and periodically issues up-to-date bibliographies of new acquisitions)

The Architectural Index
Boulder, Colo.
Annual index to articles appearing in *Architecture*, *Architectural Record*, *Housing*, *Interior Design*, *Interiors*, *Journal of Architectural Research*, *Landscape Architecture*, *Progressive Architecture*, *Research and Design*, and *Residential Interiors*.
(helpful in locating articles that appear in magazines and journals; entries arranged by building type, architect, and location)

Avery Architectural and Fine Arts Library
Columbia University
New York, N.Y. 10017
(one of the finest architectural libraries in the United States; contains the original drawings of many prominent architects)

Bettman Archive
136 East 57th Street
New York, N.Y. 10022
(a commercial source of pictorial materials; can be helpful in finding specific historic documentation)

Cooperative Preservation of Architectural Records (COPAR)
Prints and Photographs Division
Library of Congress
Washington, D.C. 20540
(address requests for information to Curator, Architecture, Design and Engineering Collections)

Prairie Archives
Milwaukee Art Center
750 North Lincoln Memorial Drive
Milwaukee, Wis. 53202
(along with The Art Institute of Chicago, the primary archive of Midwest architecture)

EDUCATION AND TRAINING

ANNUAL SEMINARS

American Association of Museums
1055 Thomas Jefferson Street, N.W.
Washington, D.C.20007

American Institute of Architects
1735 New York Avenue, N.W.
Washington, D.C. 20006

Association for Preservation Technology
P.O. Box 2487, Station D
Ottawa, Ontario K1P 5W6, Canada

Practising Law Institute
810 Seventh Avenue
New York, N.Y. 10019

Preservation Action
1700 Connecticut Avenue, N.W.
Washington, D.C. 20009

Society for Commercial Archaeology
National Museum of American History
Washington, D.C. 20560

Society for Industrial Archaeology
National Museum of American History
Washington, D.C. 20560

Society for the Preservation of New England Antiquities
141 Cambridge Street
Boston, Mass. 02114

Society of Architectural Historians
1700 Walnut Street
Philadelphia, Pa. 19103

The Victorian Society in America
219 South Sixth Street
Philadelphia, Pa. 19106

SUMMER WORKSHOPS

Cornell University Summer Institute on Historic Preservation Planning
209 West Sibley Hall
Ithaca, N.Y. 14853
(one week, for professionals and nonprofessionals)

Harvard Graduate School of Design
Office of Special Programs
48 Quincy Street
Cambridge, Mass. 02138
(various summer and regular courses, for professionals and students)

Preservation Institute: Nantucket
c/o Department of Agriculture
University of Florida
Gainesville, Fla. 32611
(nine weeks, for graduate students and advanced undergraduates)

Seminars on American Culture
New York State Historical Association
Lake Road, Route 80
Cooperstown, N.Y. 13326
(one week, open to general public)

ADDITIONAL COURSES

Campbell Center for Historic Preservation Studies
Box 66
Mount Carroll, Ill. 61053
(training in general preservation and in specific craft skills)

Center for Preservation Training
National Trust for Historic Preservation
1785 Massachusetts Avenue, N.W.
Washington, D.C. 20036
(variety of workshops and conferences on all aspects of preservation)

National Preservation Institute
c/o National Building Museum
440 G Street, N.W.
Washington, D.C. 20001
(various courses, open to general public)

Restoration Workshop
National Trust for Historic Preservation
Lyndhurst
635 South Broadway
Tarrytown, N.Y. 10591
(apprenticeship training in all aspects of restoration)

RESTORE
19 West 44th Street
New York, N.Y. 10036
(thirty weeks, for contractors, craftsmen, and others interested in masonry preservation and restoration)

INDEX

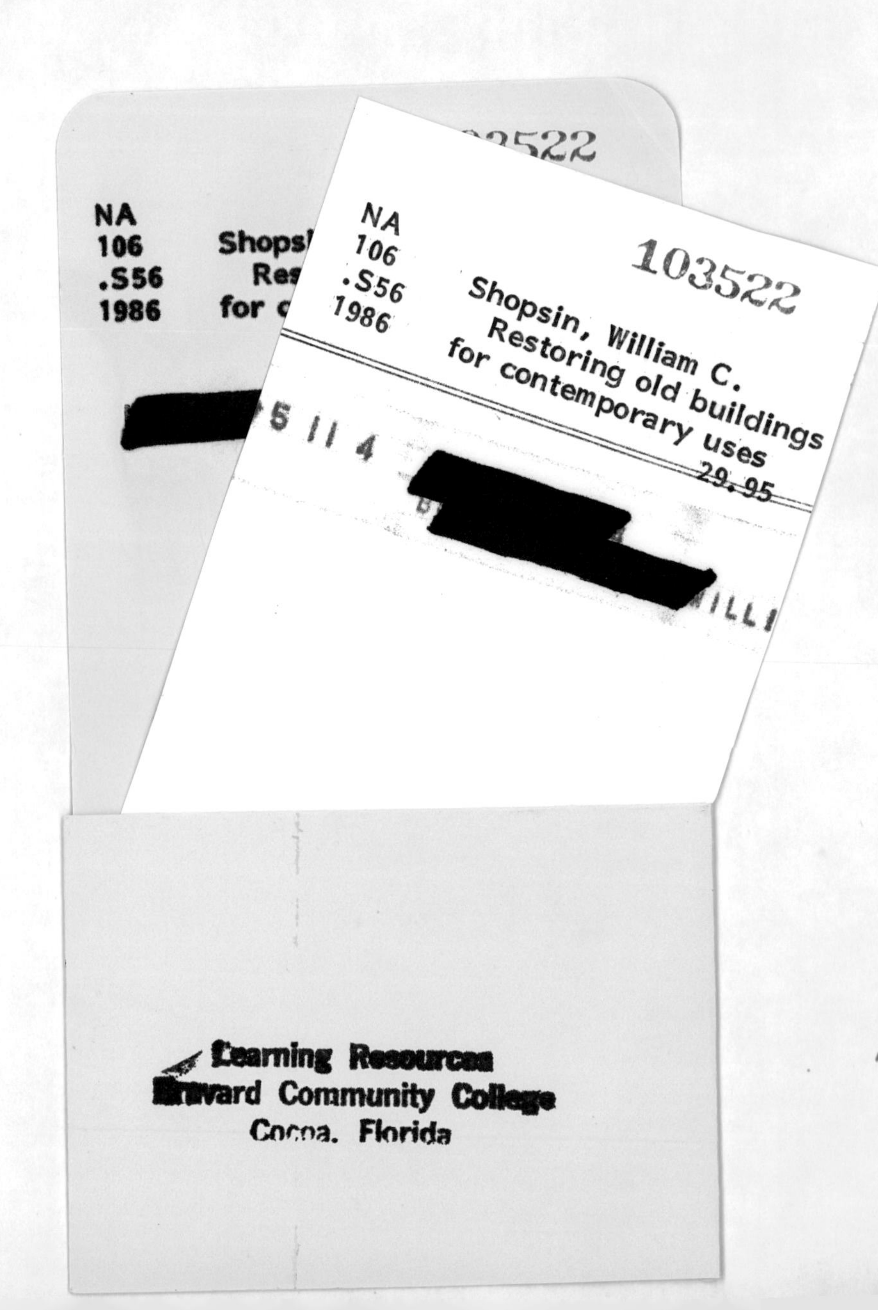
NA
106
.S56
1986
103522
Shopsin, William C.
Restoring old buildings
for contemporary uses
29.95
Learning Resources
Brevard Community College
Cocoa, Florida